UNDERGRADUATE
CONVEXITY
From Fourier and Motzkin to Kuhn and Tucker

UNDERGRADUATE CONVEXITY

From Fourier and Motzkin to Kuhn and Tucker

Niels Lauritzen

Aarhus University, Denmark

World Scientific

NEW JERSEY · LONDON · SINGAPORE · BEIJING · SHANGHAI · HONG KONG · TAIPEI · CHENNAI

Published by

World Scientific Publishing Co. Pte. Ltd.

5 Toh Tuck Link, Singapore 596224

USA office: 27 Warren Street, Suite 401-402, Hackensack, NJ 07601

UK office: 57 Shelton Street, Covent Garden, London WC2H 9HE

British Library Cataloguing-in-Publication Data
A catalogue record for this book is available from the British Library.

Cover image: Johan Ludvig William Valdemar Jensen (1859–1925).
Mathematician and telephone engineer.
Photograph by Vilhelm Rieger (courtesy of the Royal Library, Copenhagen).

UNDERGRADUATE CONVEXITY
From Fourier and Motzkin to Kuhn and Tucker

ISBN 978-981-4412-51-3
ISBN 978-981-4452-76-2 (pbk)

Printed in Singapore by B & Jo Enterprise Pte Ltd

Preface

Convexity is a key concept in modern mathematics with rich applications in economics and optimization.

This book is a basic introduction to convexity based on several years of teaching the one-quarter courses *Konvekse Mœngder* (convex sets) and *Konvekse Funktioner* (convex functions) to undergraduate students in mathematics, economics and computer science at Aarhus University. The prerequisites are minimal consisting only of first year courses in calculus and linear algebra.

I have attempted to strike a balance between different approaches to convexity in applied and pure mathematics. Compared to the former the mathematics takes a front seat. Compared to some of the latter, a key point is that the ability to carry out computations is considered paramount and a crucial stepping stone to the understanding of abstract concepts e.g., the definition of a face of a convex set does not make much sense before it is viewed in the context of several simple examples and computations.

Chapters 1–6 treat convex subsets from the basics of linear inequalities to Minkowski's theorem on separation of disjoint convex subsets by hyperplanes. The basic idea has been to emphasize part of the rich "finite" theory of polyhedra before entering into the "infinite" theory of closed convex subsets.

Fourier-Motzkin elimination is to linear inequalities what Gaussian elimination is to linear equations. It seems appropriate to begin a course on convexity by introducing this simple, yet powerful method. The prerequisites are barely present. Still the first chapter contains substantial results such as a simple algorithm for linear optimization and the fundamental theorem that projections of polyhedra are themselves polyhedra.

Before introducing closed convex subsets, several basic definitions and highlights from the polyhedral world are given: a concise treatment of affine subspaces, faces of convex subsets, Bland's rule from the simplex algorithm as a tool for computing with the convex hull, faces of polyhedra, Farkas's lemma, steady states for Markov chains, duality in linear programming, doubly stochastic matrices and the Birkhoff polytope.

The chapter *Computations with polyhedra* contains a treatment of two important polyhedral algorithms: the double description method and the simplex algorithm. The double description method is related to Fourier-Motzkin elimination. It is very easily explained in an undergraduate context especially as a vehicle for computing the bounding half spaces of a convex hull.

The simplex algorithm solves linear optimization problems and is somewhat mysterious from a mathematical perspective. There is no obvious reason it should work well. In fact, the famous mathematician John von Neumann never really believed it would perform in practice. The inventor George B. Dantzig also searched for alternate methods for years before confronting experimental data from some of the world's first computers: the simplex algorithm performed amazingly well in practice. Only recently has a mathematical explanation for this phenomenon been given by Spielman and Teng. Our treatment of the simplex algorithm and the simplex tableau deviates from the standard form and works with the polyhedron in its defining space.

The transition to the "continuous" theory of non-polyhedral convex subsets comes after the first five chapters. Here it is proved that closed convex subsets serve as generalizations of polyhedra, since they coincide with arbitrary intersections of affine half spaces. The existence of a supporting hyperplane at a boundary point of a convex subset is proved and Minkowski's theorems on compact convex subsets and separation of disjoint convex subsets are given.

Chapters 7–10 treat convex functions from the basic theory of convex functions of one variable with Jensen's inequality to the Karush-Kuhn-Tucker conditions, dual optimization problems and an outline of an interior point algorithm for solving convex optimization problems in several variables. The setting is almost always the simplest. Great generality is fine when you have lived with a subject for years, but in an introductory course it tends to become a burden. You accomplish less by including more.

The main emphasis is on differentiable convex functions. Since undergraduate knowledge of differentiability may vary, we give an almost complete review of the theory of differentiability in one and several variables. The only "general" result on convex functions not assuming differentiability is the existence of the subgradient at a point.

An understanding of convex functions of several variables is impossible without knowledge of the finer points of linear algebra over the real numbers. Introducing convex functions of several variables, we also give a thorough review of positive semidefinite matrices and reduction of symmetric matrices. This important part of linear algebra is rarely fully understood at an undergraduate level.

The final chapter treats *Convex optimization*. The key elements are the Karush-Kuhn-Tucker conditions, how saddle points of the Lagrangian lead to a dual optimization problem and finally an outline of an interior point algorithm using bisection and the modified Newton method. Monographs have been written on these three topics. We only give a brief but self-contained introduction with simple examples.

Suggestions for teaching a one-semester course

The amount of material included in this book exceeds a realistic plan for a one-semester undergraduate course on convexity. I consider Fourier-Motzkin elimination (Chapter 1), affine subspaces (Chapter 2), basics of convex subsets (Chapter 3), the foundational material on polyhedra in Chapter 4, a taste of one of the two algorithms in Chapter 5 and closed convex subsets (Chapter 6) as minimum along with almost all of the material in Chapters 7–10.

The progression of learning depends on the proficiency in linear algebra and calculus. The necessary basic concepts from analysis are introduced in Appendix A. In Appendix B there is a review of linear algebra from the point of view of linear equations leading to the rank of a matrix.

In my view, a too rigid focus on the abstract mathematical details before telling about examples and computations is a major setback in the teaching of mathematics at all levels. Certainly the material in this book benefits from being presented in a computational context with lots of examples.

Aarhus, December 2012

Acknowledgments

I am extremely grateful to Tage Bai Andersen and Jesper Funch Thomsen for very useful and detailed comments on a second draft for this book. Comments from Kent Andersen, Jens Carsten Jantzen, Anders Nedergaard Jensen and Markus Kiderlen also led to several improvements.

I am an algebraist by training and encountered convexity because of an interest in computational algebra (and computers!). As such, I have benefited immensely over the years from insightful explanations from the following more knowledgeable people: Tage Bai Andersen, Kent Andersen, Kristoffer Arnsfelt Hansen, Peter Bro Miltersen, Marcel Bökstedt, Komei Fukuda, Anders Nedergaard Jensen, Herbert Scarf, Jacob Schach Møller, Andrew du Plessis, Henrik Stetkær, Bernd Sturmfels, Rekha Thomas, Jørgen Tornehave, Jørgen Vesterstrøm and Bent Ørsted.

I am grateful to Jens Carsten Jantzen, Jesper Lützen and Tage Gutmann Madsen for help in tracking down the venerable Jensen inequality postage stamp used for several years by the Department of Mathematical Sciences at University of Copenhagen. Also, thanks to Tinne Hoff Kjeldsen for sharing her expertise on the fascinating history of convexity and optimization.

A very special thanks to the teaching assistants on *Konvekse Mængder* and *Konvekse Funktioner*: Lisbeth Laursen, Jonas Andersen Seebach, Morten Leander Petersen, Rolf Wognsen, Linnea Jørgensen and Dan Zhang. They pointed out several inaccuracies in my lecture notes along the way.

I am grateful to Kwong Lai Fun and Lakshmi Narayanan of World Scientific for their skilled help in the production of this book.

Lars 'daleif' Madsen has been crucial in the technical typesetting with his vast knowledge of LaTeX and his usual careful attention to detail.

Finally, Helle and William deserve an abundance of gratitude for their patience and genuine love.

Contents

Chapter 1

Fourier-Motzkin elimination

You probably agree that it is easy to solve the equation

$$2x = 4. \tag{1.1}$$

This is an example of a linear equation in one variable having the unique solution $x = 2$. Perhaps you will be surprised to learn, that there is essentially no difference between solving a simple equation like (1.1) and the more complicated system

$$\begin{aligned}
2x + y + z &= 7 \\
x + 2y + z &= 8 \\
x + y + 2z &= 9
\end{aligned} \tag{1.2}$$

of linear equations in x, y and z. Using the first equation $2x + y + z = 7$ we solve for x and get

$$x = (7 - y - z)/2. \tag{1.3}$$

This may be substituted into the remaining two equations in (1.2) and we get the simpler system

$$\begin{aligned}
3y + z &= 9 \\
y + 3z &= 11
\end{aligned}$$

of linear equations in y and z. Again using the first equation in this system we get

$$y = (9 - z)/3 \tag{1.4}$$

ending up with the simple equation $8z = 24$. This is an equation of the type in (1.1) giving $z = 3$. Now $z = 3$ gives $y = 2$ using (1.4). Finally $y = 2$ and $z = 3$ gives $x = 1$ using (1.3).

Figure 1.1: Isaac Newton (1642–1727). English mathematician.

Solving a seemingly complicated system of linear equations like (1.2) is really no more difficult than solving the simple equation (1.1). One of the world's greatest scientists, Isaac Newton, found it worthwhile to record this method in 1720 with the words

> *And you are to know, that by each Æquation one unknown Quantity may be taken away, and consequently, when there are as many Æquations and unknown Quantities, all at length may be reduc'd into one, in which there shall be only one Quantity unknown.*

Figure 1.2: Carl Friedrich Gauss (1777–1855). German mathematician.

During the computation of the orbit of the asteroid Pallas around 1810, Gauss encountered the need for solving linear equations related to his famous least squares method. If you spend a little time deciphering the Latin in Gauss's original writings (see Figure 1.3), you will see how elimination appears naturally towards the end of the page. In spite of Newton's explicit description several years before Gauss was born, this procedure is now known as Gaussian elimination (see [Grcar (2011)] for more on the fascinating history of Gaussian elimination).

20 DISQUISITIO

13.

Quum sex incognitas dL, $d7$ etc. ita determinare non liceat, ut omnibus undecim aequationibus exacte satisfiat, i. e. ut singulae incognitarum functiones quae sunt ad dextram simul fiant $= 0$, valores eos eruemus, per quos functionum harum quadrata summam quam minimam efficiant. Facile quidem perspicitur, si generaliter functiones lineares incognitarum p, q, r, s etc. propositae sint hae

$$\begin{aligned} n + ap + bq + cr + ds + \text{etc.} &= w \\ n' + a'p + b'q + c'r + d's + \text{etc.} &= w' \\ n'' + a''p + b''q + c''r + d''s + \text{etc.} &= w'' \\ n''' + a'''p + b'''q + c'''r + d'''s + \text{etc.} &= w''' \end{aligned}$$

etc., aequationes conditionales, ut $ww + w'w' + w''w'' + w'''w''' + \text{etc.} = \Omega$ fiat minimum, esse hasce

$$\begin{aligned} aw + a'w' + a''w'' + a'''w''' + \text{etc.} &= 0 \\ bw + b'w' + b''w'' + b'''w''' + \text{etc.} &= 0 \\ cw + c'w' + c''w'' + c'''w''' + \text{etc.} &= 0 \\ dw + d'w' + d''w'' + d'''w''' + \text{etc.} &= 0 \end{aligned}$$

etc. sive

designando brevitatis causa

$$\begin{aligned} &an + a'n' + a''n'' + a'''n''' + \text{etc.} \quad \text{per } [an] \\ &aa + a'a' + a''a'' + a'''a''' + \text{etc.} \quad \text{per } [aa] \\ &ab + a'b' + a''b'' + a'''b''' + \text{etc.} \quad \text{per } [ab] \\ &\text{etc.} \\ &bb + b'b' + b''b'' + b'''b''' + \text{etc.} \quad \text{per } [bb] \\ &bc + b'c' + b''c'' + b'''c''' + \text{etc.} \quad \text{per } [bc] \\ &\text{etc. etc.} \end{aligned}$$

p, q, r, s etc. determinari debere per eliminationem ex aequationibus

$$\begin{aligned} [an] + [aa]p + [ab]q + [ac]r + [ad]s + \text{etc.} &= 0 \\ [bn] + [ab]p + [bb]q + [bc]r + [bd]s + \text{etc.} &= 0 \\ [cn] + [ac]p + [bc]q + [cc]r + [cd]s + \text{etc.} &= 0 \\ [dn] + [ad]p + [bd]q + [cd]r + [dd]s + \text{etc.} &= 0 \end{aligned}$$

etc.

Figure 1.3: Gauss's encounter with (Gaussian) elimination (published in 1810) for use in the least squares method in computing the orbit of the asteroid Pallas. Notice that Ω is the sum of squares to be minimized.

1.1 Linear inequalities

Inequalities may be viewed as a generalization of equations, since for two numbers a and b, $a = b$ if and only if $a \leq b$ and $a \geq b$. We will describe

Figure 1.4: Joseph Fourier (1768–1830). French mathematician.

a clever algorithm going back to Fourier for systematically solving systems of linear inequalities. Fourier had a rather concrete problem in mind, when he presented his note [Fourier (1826)] (see Figure 1.6).

The algorithm itself is very similar to Gaussian elimination except that we are facing two types of inequalities ($\leq$) and ($\geq$) instead of just one equality ($=$). First consider the simplest case of linear inequalities in just one variable by way of the example

$$
\begin{aligned}
2x + 1 &\leq 7 \\
3x - 2 &\leq 4 \\
-x + 2 &\leq 3 \\
x &\geq 0
\end{aligned}
\tag{1.5}
$$

of inequalities in x. This can be rewritten to

$$
\begin{aligned}
x &\leq 3 \\
x &\leq 2 \\
-1 &\leq x \\
0 &\leq x
\end{aligned}
$$

and therefore

$$
\begin{aligned}
x &\leq \min\{2,3\} = 2 \\
\max\{-1,0\} = 0 &\leq x
\end{aligned}
$$

or simply $0 \leq x \leq 2$. Here the fundamental difference between linear equations and linear inequalities is apparent. Multiplying by -1 leaves $=$ invariant, whereas $\leq$ changes into $\geq$.

Consider now the system

$$
\begin{aligned}
x & & &\geq 0 \\
x &+ 2y &&\leq 6 \\
x &+ y &&\geq 2 \\
x &- y &&\leq 3 \\
& y &&\geq 0
\end{aligned}
\tag{1.6}
$$

of inequalities in two variables x and y.

Perhaps the most straightforward way of approaching (1.6) is through a sketch. The bounding lines are

$$
\begin{aligned}
x & & &= 0 \\
x &+ 2y &&= 6 \\
x &+ y &&= 2 \\
x &- y &&= 3 \\
& y &&= 0.
\end{aligned}
\tag{1.7}
$$

For each line we pick a point to decide which half plane to shade e.g., we need to shade below the line $x + 2y = 6$, since the corresponding inequality is $x + 2y \leq 6$ and (for example) $0 + 2 \cdot 0 < 6$. The intersection of these half planes is sketched as the shaded area in Figure 1.5.

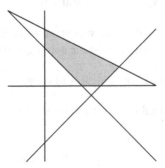

Figure 1.5: Sketch of the solutions in (1.6).

We are aiming for a more effective way of representing the solutions. Our sketching techniques are not of much use solving for example 17 linear inequalities in 12 unknowns. In order to attack (1.6) algebraically, we first record the following result strongly related to Fourier's problem in Figure 1.6.

Proposition 1.1. *Let $\alpha_1, \ldots, \alpha_r, \beta_1, \ldots, \beta_s \in \mathbb{R}$. Then*

$$\max\{\alpha_1, \ldots, \alpha_r\} \leq \min\{\beta_1, \ldots, \beta_s\}$$

if and only if $\alpha_i \leq \beta_j$ for every i, j with $1 \leq i \leq r$ and $1 \leq j \leq s$:

$$\alpha_1 \leq \beta_1 \ldots \alpha_1 \leq \beta_s$$
$$\vdots \qquad \ddots \qquad \vdots$$
$$\alpha_r \leq \beta_1 \ldots \alpha_r \leq \beta_s.$$

Proof.   If $\max\{\alpha_1, \ldots, \alpha_r\} \leq \min\{\beta_1, \ldots, \beta_s\}$, then

$$\alpha_i \leq \max\{\alpha_1, \ldots, \alpha_r\} \leq \min\{\beta_1, \ldots, \beta_s\} \leq \beta_j$$

for every $1 \leq i \leq r$ and $1 \leq j \leq s$. On the other hand let $1 \leq i_0 \leq r$ and $1 \leq j_0 \leq s$ be such that

$$\alpha_{i_0} = \max\{\alpha_1, \ldots, \alpha_r\}$$
$$\beta_{j_0} = \min\{\beta_1, \ldots, \beta_s\}.$$

If $\alpha_i \leq \beta_j$ for every $1 \leq i \leq r$ and $1 \leq j \leq s$, then $\alpha_{i_0} \leq \beta_{j_0}$. Therefore

$$\max\{\alpha_1, \ldots, \alpha_r\} \leq \min\{\beta_1, \ldots, \beta_s\}. \qquad \square$$

Inspired by Gaussian elimination we will attempt to isolate and eliminate x. The key point here is that there exists x solving the two inequalities

$$a \leq x \quad \text{and} \quad x \leq b \quad \text{if and only if} \quad a \leq b,$$

where a and b are real numbers. With this in mind, we rewrite (1.6) to

$$\begin{aligned} 0 &\leq x \\ x &\leq 6 - 2y \\ 2 - y &\leq x \\ x &\leq 3 + y \\ y &\geq 0. \end{aligned}$$

Just like in one variable, this system can be reduced to

$$\begin{aligned} x &\leq \min\{6 - 2y, 3 + y\} \\ \max\{0, 2 - y\} &\leq x \\ y &\geq 0. \end{aligned} \qquad (1.8)$$

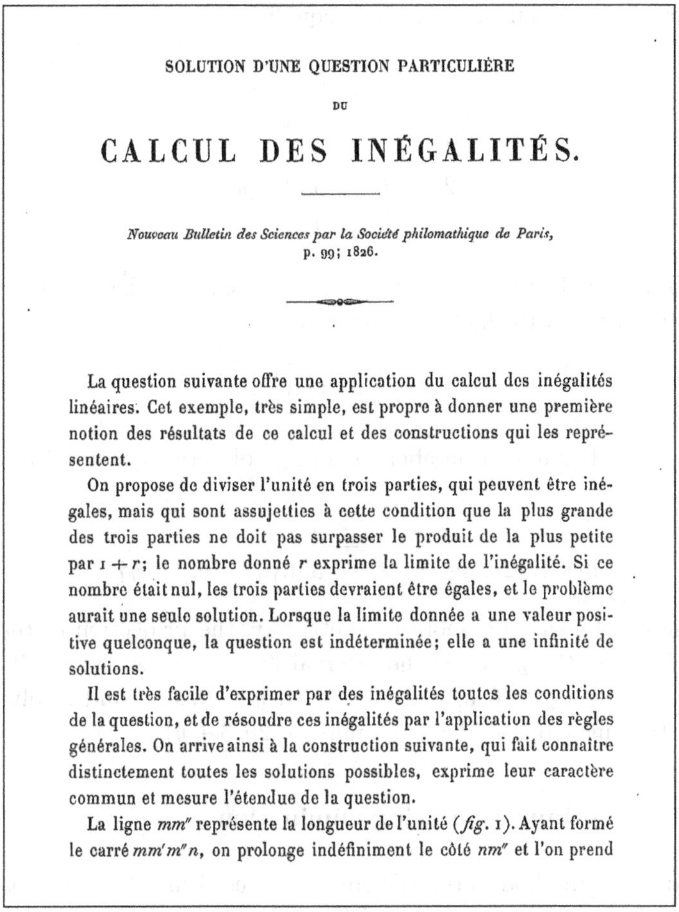

SOLUTION D'UNE QUESTION PARTICULIÈRE

DU

CALCUL DES INÉGALITÉS.

Nouveau Bulletin des Sciences par la Société philomathique de Paris,
p. 99; 1826.

La question suivante offre une application du calcul des inégalités linéaires. Cet exemple, très simple, est propre à donner une première notion des résultats de ce calcul et des constructions qui les représentent.

On propose de diviser l'unité en trois parties, qui peuvent être inégales, mais qui sont assujetties à cette condition que la plus grande des trois parties ne doit pas surpasser le produit de la plus petite par 1 + r; le nombre donné r exprime la limite de l'inégalité. Si ce nombre était nul, les trois parties devraient être égales, et le problème aurait une seule solution. Lorsque la limite donnée a une valeur positive quelconque, la question est indéterminée; elle a une infinité de solutions.

Il est très facile d'exprimer par des inégalités toutes les conditions de la question, et de résoudre ces inégalités par l'application des règles générales. On arrive ainsi à la construction suivante, qui fait connaître distinctement toutes les solutions possibles, exprime leur caractère commun et mesure l'étendue de la question.

La ligne *mm″* représente la longueur de l'unité (*fig.* 1). Ayant formé le carré *mm′m″n*, on prolonge indéfiniment le côté *nm″* et l'on prend

Figure 1.6: The first page of Fourier's note [Fourier (1826)]. Notice the specific problem he is describing. In modern parlor it amounts to finding all x, y, z such that $x + y + z = 1$ and $\max\{x, y, z\} \leq (1 + r) \min\{x, y, z\}$ for a fixed $r \geq 0$.

Therefore we can eliminate x from (1.8) and deduce that

$$\begin{aligned} \max\{0, 2 - y\} &\leq \min\{6 - 2y, 3 + y\} \\ y &\geq 0 \end{aligned} \tag{1.9}$$

is solvable in y if and only if (1.8) is solvable in x and y. Now Proposition 1.1

shows that (1.9) is equivalent to the inequalities

$$
\begin{aligned}
0 &\leq 6 - 2y \\
0 &\leq 3 + y \\
2 - y &\leq 6 - 2y \\
2 - y &\leq 3 + y \\
0 &\leq y
\end{aligned}
$$

in the variable y. These inequalities can be solved just like we solved (1.5) and may be reduced to the two inequalities

$$0 \leq y \leq 3.$$

We have proved that two numbers x and y solve the system (1.6) if and only if

$$
\begin{aligned}
0 \leq y &\leq 3 \\
\max\{0, 2 - y\} \leq x &\leq \min\{6 - 2y, 3 + y\}.
\end{aligned}
$$

If you phrase things a bit more geometrically, the projection of the solutions to (1.6) on the y-axis is the interval $[0, 3]$. In other words, if x and y solve (1.6), then $y \in [0, 3]$ and for a fixed $y \in [0, 3]$, x and y solve (1.6) provided that $\max\{0, 2 - y\} \leq x \leq \min\{6 - 2y, 3 + y\}$.

1.2   Linear optimization using elimination

The elimination method outlined here can be used in solving the problem of maximizing a linear function subject to constraints consisting of linear inequalities (see also Exercises 1.5 and 1.9). Such linear optimization problems are excellent models for many practical problems and are usually solved with the more advanced simplex algorithm, which we will explain later in §5.4. The following example illustrates how elimination is used by adjoining an extra variable for the linear function.

Example 1.2. Find the maximal value of $x + y$ subject to the constraints

$$
\begin{aligned}
x + 2y &\leq 3 \\
2x + y &\leq 3 \\
x &\geq 0 \\
y &\geq 0.
\end{aligned}
$$

Here the trick is to introduce an extra variable z and then find the maximal z, such that the inequalities (and the one equation)

$$
\begin{aligned}
x + y &= z \\
x + 2y &\leq 3 \\
2x + y &\leq 3 \\
x &\geq 0 \\
y &\geq 0
\end{aligned}
$$

have a solution. First we eliminate x by substituting $x = z - y$ into the inequalities:

$$
\begin{aligned}
z + y &\leq 3 \\
2z - y &\leq 3 \\
z - y &\geq 0 \\
y &\geq 0.
\end{aligned}
$$

Preparing for elimination of y we write

$$
\begin{aligned}
y &\leq 3 - z \\
2z - 3 &\leq y \\
y &\leq z \\
0 &\leq y.
\end{aligned}
$$

Therefore

$$
\max\{0, 2z - 3\} \leq y \leq \min\{3 - z, z\} \tag{1.10}
$$

and Proposition 1.1 applies to give the inequalities

$$
\begin{aligned}
0 &\leq 3 - z \\
0 &\leq z \\
2z - 3 &\leq 3 - z \\
2z - 3 &\leq z
\end{aligned}
$$

with solution $0 \leq z \leq 2$. Therefore the maximal value of $z = x + y$ is 2. You can obtain a solution (x, y) to the linear optimization problem by first inserting $z = 2$ into (1.10) to get y and then insert z and y into $x = z - y$ to get x. This gives the unique optimum $(x, y) = (1, 1)$.

Now we are ready to enter into the general setting.

1.3 Polyhedra

A *linear inequality* in n variables $x_1, \ldots, x_n$ is an inequality of the form

$$
a_1 x_1 + \cdots + a_n x_n \leq b,
$$

where $a_1, \ldots, a_n, b \in \mathbb{R}$. By $\mathbb{R}^n$ we denote the set of *column vectors* with n entries ($n \times 1$ matrices). For typographical reasons we will sometimes let $(x_1, \ldots, x_n)$ refer to the column vector in $\mathbb{R}^n$ with entries $x_1, \ldots, x_n$.

Definition 1.3. The subset

$$P = \left\{ \begin{pmatrix} x_1 \\ \vdots \\ x_n \end{pmatrix} \in \mathbb{R}^n \; \middle| \; \begin{array}{c} a_{11}x_1 + \cdots + a_{1n}x_n \leq b_1 \\ \vdots \\ a_{m1}x_1 + \cdots + a_{mn}x_n \leq b_m \end{array} \right\} \subseteq \mathbb{R}^n$$

of solutions to a system

$$
\begin{array}{ccccc}
a_{11}x_1 & + \cdots + & a_{1n}x_n & \leq & b_1 \\
& & & \vdots & \\
a_{m1}x_1 & + \cdots + & a_{mn}x_n & \leq & b_m
\end{array}
$$

of finitely many linear inequalities (here a_{ij} and b_i are real numbers) is called a *polyhedron*.

Notation 1.4. For vectors $u = (u_1, \ldots, u_n)$, $v = (v_1, \ldots, v_n) \in \mathbb{R}^n$ we introduce the notation

$$u \leq v \quad \Longleftrightarrow \quad u_1 \leq v_1 \text{ and } \ldots \text{ and } u_n \leq v_n.$$

With this convention the polyhedron in Definition 1.3 is expressed more economically as

$$P = \{x \in \mathbb{R}^n \,|\, Ax \leq b\},$$

where A is the $m \times n$ matrix and b the vector in $\mathbb{R}^m$ given by

$$A = \begin{pmatrix} a_{11} & \cdots & a_{1n} \\ \vdots & \ddots & \vdots \\ a_{m1} & \cdots & a_{mn} \end{pmatrix} \quad \text{and} \quad b = \begin{pmatrix} b_1 \\ \vdots \\ b_m \end{pmatrix}.$$

Example 1.5. For $u = (1, 2)$ and $v = (2, 3)$, $u \leq v$, whereas neither $u \leq v$ nor $v \leq u$ hold for $u = (1, 0)$ and $v = (0, 1)$. The polyhedron in (1.6) can be written as

$$P = \left\{ \begin{pmatrix} x \\ y \end{pmatrix} \in \mathbb{R}^2 \; \middle| \; \begin{pmatrix} -1 & 0 \\ 1 & 2 \\ -1 & -1 \\ 1 & -1 \\ 0 & -1 \end{pmatrix} \begin{pmatrix} x \\ y \end{pmatrix} \leq \begin{pmatrix} 0 \\ 6 \\ -2 \\ 3 \\ 0 \end{pmatrix} \right\}.$$

In modern mathematical terms our computations can be used in proving the main result (Theorem 1.6) in this chapter that the projection of a polyhedron is a polyhedron (see Figure 1.7). This seemingly innocuous result has rather profound consequences. The proof may appear a bit technical at first, but it is simply a formalization of the concrete computations in §1.1.

The elimination method for linear inequalities in §1.1 is called *Fourier-Motzkin elimination*. Not knowing the classical paper by Fourier, Motzkin[1] rediscovered it in his thesis "Beiträge zur Theorie der linearen Ungleichungen" supervised by Ostrowski[2] in Basel, 1933.

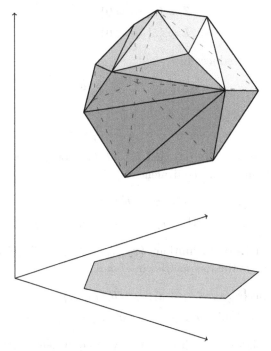

Figure 1.7: Projection to $\mathbb{R}^2$ of a polyhedron in $\mathbb{R}^3$.

Theorem 1.6. *Consider the projection* $\pi : \mathbb{R}^n \to \mathbb{R}^{n-1}$ *given by*

$$\pi(x_1, \ldots, x_n) = (x_2, \ldots, x_n).$$

If $P \subseteq \mathbb{R}^n$ *is a polyhedron, then*

$$\pi(P) = \{(x_2, \ldots, x_n) \mid \exists\, x_1 \in \mathbb{R} : (x_1, x_2, \ldots, x_n) \in P\} \subseteq \mathbb{R}^{n-1}$$

---

[1]Theodore Samuel Motzkin (1908–1970). Israeli-American mathematician.
[2]Alexander Markowich Ostrowski (1893–1986). Russian-Swiss mathematician.

is a polyhedron.

Proof.   Suppose that P is the set of solutions to

$$a_{11}x_1 + \cdots + a_{1n}x_n \leq b_1$$
$$\vdots$$
$$a_{m1}x_1 + \cdots + a_{mn}x_n \leq b_m \,.$$

We partition these m inequalities according to the sign of a_{i1}:

$$G = \{i \,|\, a_{i1} > 0\}$$
$$Z = \{i \,|\, a_{i1} = 0\}$$
$$L = \{i \,|\, a_{i1} < 0\}.$$

Inequality number i reduces to

$$x_1 \leq a'_{i2}x_2 + \cdots + a'_{in}x_n + b'_i,$$

if $i \in G$ and to

$$a'_{j2}x_2 + \cdots + a'_{jn}x_n + b'_j \leq x_1,$$

if $j \in L$, where $a'_{ik} = -a_{ik}/a_{i1}$ and $b'_i = b_i/a_{i1}$ for $k = 2, \ldots, n$. So the inequalities in L and G are equivalent to

$$\max\{a'_{i2}x_2 + \cdots + a'_{in}x_n + b'_i \,|\, i \in L\}$$
$$\leq x_1 \leq \min\{a'_{j2}x_2 + \cdots + a'_{jn}x_n + b'_j \,|\, j \in G\}$$

by Proposition 1.1. By definition, $(x_2, \ldots, x_n) \in \pi(P)$ if and only if $(x_2, \ldots, x_n)$ satisfies the inequalities in Z and

$$\max\{a'_{i2}x_2 + \cdots + a'_{in}x_n + b'_i \,|\, i \in L\}$$
$$\leq \min\{a'_{j2}x_2 + \cdots + a'_{jn}x_n + b'_j \,|\, j \in G\}.$$

Proposition 1.1 shows that this inequality is equivalent to the $|L||G|$ inequalities in $x_2, \ldots, x_n$ consisting of

$$a'_{i2}x_2 + \cdots + a'_{in}x_n + b'_i \leq a'_{j2}x_2 + \cdots + a'_{jn}x_n + b'_j$$

or rather

$$(a'_{i2} - a'_{j2})x_2 + \cdots + (a'_{in} - a'_{jn})x_n \leq b'_j - b'_i$$

where $i \in L$ and $j \in G$. Adding the inequalities in Z, where x_1 is not present, it follows that $\pi(P)$ is the set of solutions to these $|L||G| + |Z|$ linear inequalities in $x_2, \ldots, x_n$. Therefore $\pi(P)$ is a polyhedron. □

1.4 Exercises

Exercise 1.1. Sketch the set of solutions to the system

$$
\begin{aligned}
2x + y &\geq 2 \\
3x + y &\leq 9 \\
-x + 2y &\leq 4 \\
y &\geq 0
\end{aligned}
\tag{1.11}
$$

of linear inequalities. Carry out the elimination procedure for (1.11) as illustrated in §1.1.

Exercise 1.2. Let

$$
P = \left\{ (x, y, z) \in \mathbb{R}^3 \;\middle|\;
\begin{aligned}
-x - y - z &\leq 0 \\
3x - y - z &\leq 1 \\
-x + 3y - z &\leq 2 \\
-x - y + 3z &\leq 3
\end{aligned}
\right\}
$$

and $\pi : \mathbb{R}^3 \to \mathbb{R}^2$ be given by $\pi(x, y, z) = (y, z)$.

   (i) Compute $\pi(P)$ as a polyhedron i.e., as the solutions to a set of linear inequalities in y and z.
   (ii) Compute $\eta(P)$, where $\eta : \mathbb{R}^3 \to \mathbb{R}$ is given by $\eta(x, y, z) = x$.
   (iii) How many integral points does P contain i.e., how many elements are in the set $P \cap \mathbb{Z}^3$?

Exercise 1.3. Find all solutions $x, y, z \in \mathbb{Z}$ to the linear inequalities

$$
\begin{aligned}
-x + y - z &\leq 0 \\
- y + z &\leq 0 \\
- z &\leq 0 \\
x \qquad - z &\leq 1 \\
y \qquad &\leq 1 \\
z &\leq 1
\end{aligned}
$$

by using Fourier-Motzkin elimination.

Exercise 1.4. Does the system

$$
\begin{aligned}
2x - 3y + z &\leq -2 \\
x + 3y + z &\leq -3 \\
-2x - 3y + z &\leq -2 \\
-x - 3y - 3z &\leq 1 \\
-2x - y + 3z &\leq 3
\end{aligned}
$$

of linear inequalities have a solution $x, y, z \in \mathbb{R}$?

Exercise 1.5. Let $P \subseteq \mathbb{R}^n$ be a polyhedron and $c \in \mathbb{R}^n$. Define the polyhedron $P' \subseteq \mathbb{R}^{n+1}$ by

$$P' = \left\{ \begin{pmatrix} x \\ z \end{pmatrix} \in \mathbb{R}^{n+1} \, \middle| \, c^t x = z, \ x \in P, \ z \in \mathbb{R} \right\}.$$

(i) How does this setup relate to Example 1.2?

(ii) Show how projection onto the z-coordinate (and Fourier-Motzkin elimination) in P' can be used to solve the linear optimization problem of finding $x \in P$, such that $c^t x$ is minimal (or proving that such an x does not exist).

(iii) Let P denote the polyhedron from Exercise 1.2. You can see that

$$(0, 0, 0), \ (-1, \tfrac{1}{2}, \tfrac{1}{2}) \in P$$

have values 0 and -1 on their first coordinates, but what is the minimal first coordinate of a point in P?

Exercise 1.6. Solve the problem appearing in Fourier's article (Figure 1.6) for $r = 1$ using Fourier-Motzkin elimination.

Exercise 1.7. Let P denote the set of $(x, y, z) \in \mathbb{R}^3$, satisfying

$$
\begin{aligned}
-2x &+ y + z &\leq 4 \\
x & &\geq 1 \\
& y &\geq 2 \\
& z &\geq 3 \\
x &- 2y + z &\leq 1 \\
2x &+ 2y - z &\leq 5.
\end{aligned}
$$

(i) Prove that P is bounded.

(ii) Find $(x, y, z) \in P$ with z maximal. Is such a point unique?

Exercise 1.8. A vitamin pill P is produced using two ingredients M_1 and M_2. The pill needs to satisfy four constraints for the vital vitamins V_1 and V_2. It must contain at least 6 milligram and at most 15 milligram of V_1 and at least 5 milligram and at most 12 milligram of V_2. The ingredient M_1 contains 3 milligram of V_1 and 2 milligram of V_2 per gram. The ingredient M_2 contains 2 milligram of V_1 and 3 milligram of V_2 per gram:

	V_1	V_2
M_1	3	2
M_2	2	3

Let x denote the amount of M_1 and y the amount of M_2 (measured in grams) in the production of a vitamin pill. Write down a system of linear inequalities in x and y describing the constraints above.

We want a vitamin pill of minimal weight satisfying the constraints. How many grams of M_1 and M_2 should we mix? Describe how Fourier-Motzkin elimination can be used in solving this problem.

Exercise 1.9. Use Fourier-Motzkin elimination to compute the minimal value of

$$x_1 + 2x_2 + 3x_3,$$

when x_1, x_2, x_3 satisfy

$$\begin{aligned} x_1 - 2x_2 + x_3 &= 4 \\ -x_1 + 3x_2 &= 5 \end{aligned}$$

and

$$x_1 \geq 0, \quad x_2 \geq 0, \quad x_3 \geq 0.$$

•

Chapter 2

Affine subspaces

A polyhedron is the set of solutions to a system of linear inequalities. Sets

$$\{x \in \mathbb{R}^d \mid Ax = b\} \tag{2.1}$$

of solutions to a system of linear equations are polyhedra of the simplest kind. Here A is an $m \times d$ matrix and $b \in \mathbb{R}^m$ corresponding to a system of m linear equations with d unknowns.

Recall that a *line* in $\mathbb{R}^d$ is a subset (see Figure 2.1) of the form

$$\{x + t\alpha \mid t \in \mathbb{R}\},$$

where $x \in \mathbb{R}^d$ is a vector and $\alpha \in \mathbb{R}^d \setminus \{0\}$ a non-zero *directional vector*. Two distinct points $u, v \in \mathbb{R}^d$ are contained in the unique line

$$L = \{(1 - t)u + tv \mid t \in \mathbb{R}\}. \tag{2.2}$$

Here a directional vector for L is $v - u$. It is not too difficult to check that sets of solutions to systems of linear equations such as (2.1) contain the line between any two of their points. Subsets with this intrinsic geometric property are called *affine subspaces*.

The purpose of this chapter is to give an account of affine subspaces based on systems of linear equations. The difference between affine subspaces and the usual subspaces of linear algebra is that the former do not necessarily contain the zero vector. Affine subspaces also enter into the important definition of the dimension of an arbitrary subset of $\mathbb{R}^d$ so that a point has dimension zero, a line dimension one etc.

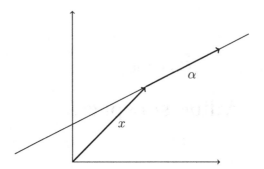

Figure 2.1: Sketch of the line $\{x + t\alpha \mid t \in \mathbb{R}\}$ in $\mathbb{R}^2$ with x and α marked.

2.1 Definition and basics

We begin by stating some basic properties of affine subspaces. First a motivating example.

Example 2.1. Consider the three points $v_1 = (2, 1, 0)$, $v_2 = (1, 0, 1)$ and $v_3 = (0, 4, -1) \in \mathbb{R}^3$. You can check through (2.2), that v_3 does not lie on the unique line through v_1 and v_2, hence there is no line containing all three points. Therefore they span a unique plane H in $\mathbb{R}^3$. This plane is given parametrically as

$$v_1 + t_1(v_2 - v_1) + t_2(v_3 - v_1) = (1 - t_1 - t_2)v_1 + t_1 v_2 + t_2 v_3 \qquad (2.3)$$

for $t_1, t_2 \in \mathbb{R}$ (see Figure 2.2). In other words, $H = v_1 + W = \{v_1 + v \mid v \in W\}$, where W is the linear subspace of $\mathbb{R}^3$ spanned by the vectors $v_2 - v_1$ and $v_3 - v_1$. With numbers inserted this reads

$$H = \left\{ \begin{pmatrix} 2 \\ 1 \\ 0 \end{pmatrix} + t_1 \begin{pmatrix} -1 \\ -1 \\ 1 \end{pmatrix} + t_2 \begin{pmatrix} -2 \\ 3 \\ -1 \end{pmatrix} \middle| t_1, t_2 \in \mathbb{R} \right\}.$$

By finding a non-zero solution to the system

$$-\alpha_1 - \alpha_2 + \alpha_3 = 0$$
$$-2\alpha_1 + 3\alpha_2 - \alpha_3 = 0$$

of linear equations, you can check that $W = \{(x, y, z) \in \mathbb{R}^3 \mid 2x + 3y + 5z = 0\}$. Therefore

$$H = \{(x, y, z) \in \mathbb{R}^3 \mid 2x + 3y + 5z = 7\} \qquad (2.4)$$

and the plane H is presented as the set of solutions to a linear equation.

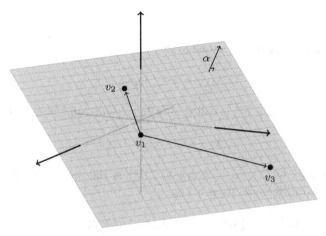

Figure 2.2: Sketch of the plane $H \subseteq \mathbb{R}^3$ marking v_1, the directional vectors $v_2 - v_1$, $v_3 - v_1$ and a normal vector α proportional to $(2, 3, 5)$.

Definition 2.2. A non-empty subset $M \subseteq \mathbb{R}^d$ is called an *affine subspace* if

$$(1 - t)u + tv \in M$$

for every $u, v \in M$ and every $t \in \mathbb{R}$. A map $f : \mathbb{R}^d \to \mathbb{R}^n$ is called an *affine map* if

$$f((1 - t)u + tv) = (1 - t)f(u) + tf(v)$$

for every $u, v \in \mathbb{R}^d$ and every $t \in \mathbb{R}$.

The identity (2.3) in Example 2.1 points to the result below.

Lemma 2.3. *Let M be an affine subspace of $\mathbb{R}^d$ and $v_1, \ldots, v_m \in M$. Then*

$$\lambda_1 v_1 + \cdots + \lambda_m v_m \in M$$

for all real numbers $\lambda_1, \ldots, \lambda_m \in \mathbb{R}$ with $\lambda_1 + \cdots + \lambda_m = 1$.

Proof. This is proved by induction on m. For $m = 2$ this is the content of Definition 2.2. For $m > 2$ we must have $1 - \lambda_i \neq 0$ for some $i = 1, \ldots, m$. We may assume that $1 - \lambda_m \neq 0$. Then

$$\lambda_1 v_1 + \cdots + \lambda_m v_m = (1 - \lambda_m)\left(\frac{\lambda_1}{1 - \lambda_m}v_1 + \cdots + \frac{\lambda_{m-1}}{1 - \lambda_m}v_{m-1}\right) + \lambda_m v_m.$$

Since $\lambda_1 + \cdots + \lambda_{m-1} = 1 - \lambda_m$ we are done by induction. $\qquad\square$

2.2   The affine hull

Definition 2.4. A linear combination

$$\lambda_1 v_1 + \cdots + \lambda_m v_m$$

of vectors $v_1, \ldots, v_m \in \mathbb{R}^d$ is called an *affine linear combination* if

$$\lambda_1 + \cdots + \lambda_m = 1.$$

The *affine hull*, $\mathrm{aff}(S)$, of a subset $S \subseteq \mathbb{R}^d$ is the set of all affine linear combinations of elements from S i.e.,

$$\mathrm{aff}(S) := \{\lambda_1 v_1 + \cdots + \lambda_m v_m \mid m \geq 1,$$
$$v_1, \ldots, v_m \in S, \lambda_1 + \cdots + \lambda_m = 1\}.$$

Proposition 2.5. *The affine hull, $\mathrm{aff}(S)$, of a subset $S \subseteq \mathbb{R}^d$ is an affine subspace. It is the smallest affine subspace containing S.*

Proof.   Suppose that $u, v \in \mathrm{aff}(S)$ i.e.,

$$u = \lambda_1 v_1 + \cdots + \lambda_r v_r$$
$$v = \mu_1 v_1' + \cdots + \mu_s v_s'$$

where $r, s \geq 1$,

$$\lambda_1 + \cdots + \lambda_r = \mu_1 + \cdots + \mu_s = 1$$

and $v_1, \ldots, v_r, v_1', \ldots, v_s' \in S$. Then

$$(1 - \alpha)u + \alpha v = (1 - \alpha)\lambda_1 v_1 + \cdots + (1 - \alpha)\lambda_r v_r + \alpha\mu_1 v_1' + \cdots + \alpha\mu_s v_s'$$

for $\alpha \in \mathbb{R}$. This is an affine linear combination, since

$$(1 - \alpha)\lambda_1 + \cdots + (1 - \alpha)\lambda_r + \alpha\mu_1 + \cdots + \alpha\mu_s$$
$$= (1 - \alpha)(\lambda_1 + \cdots + \lambda_r) + \alpha(\mu_1 + \cdots + \mu_r)$$
$$= (1 - \alpha) + \alpha = 1$$

and $\mathrm{aff}(S)$ is an affine subspace according to Definition 2.2. If M is an affine subspace containing S, then $M \supseteq \mathrm{aff}(S)$ by Lemma 2.3. This proves that $\mathrm{aff}(S)$ is the smallest subspace containing S. $\qquad\square$

2.3 Affine subspaces and subspaces

It is shown below that affine subspaces are solution sets of systems of linear equations. You may find it helpful to compare the proof with the explicit computations in Example 2.1.

Proposition 2.6. *For an affine subspace $M \subseteq \mathbb{R}^d$, $W = \{u - v \mid u, v \in M\}$ is a subspace and*

$$M = \{x_0 + w \mid w \in W\} =: x_0 + W,$$

for every $x_0 \in M$.

A subset $M \subseteq \mathbb{R}^d$ is an affine subspace if and only if it is the solution set to a system of linear equations.

If $h : \mathbb{R}^d \to \mathbb{R}^n$ is a linear map and $b \in \mathbb{R}^n$, then $f(x) = h(x) + b$ is an affine map. If $f : \mathbb{R}^d \to \mathbb{R}^n$ is an affine map, $h(x) = f(x) - f(0)$ is a linear map and $f(x) = h(x) + b$ with $b = f(0)$.

Proof. For $u_1, v_1, u_2, v_2 \in M$,

$$\lambda(u_1 - v_1) + \mu(u_2 - v_2) =$$
$$(\lambda u_1 + \mu u_2 + (1 - \lambda - \mu)v_1) - (\lambda v_1 + \mu v_2 + (1 - \lambda - \mu)v_1)$$

for $\lambda, \mu \in \mathbb{R}$ and it follows by Lemma 2.3 that $W = \{u - v \mid u, v \in M\}$ is a subspace. If $x_0 \in M$ it follows that $x_0 + (u - v) \in M$ again by Lemma 2.3. Therefore $x_0 + W \subseteq M$. On the other hand if $x \in M$, then $x = x_0 + (x - x_0) \in x_0 + W$. Therefore $M \subseteq x_0 + W$ and $M = x_0 + W$.

If M is the solution set $\{x \in \mathbb{R}^d \mid Ax = b\}$ to a system of linear equations we leave it to the reader to verify that M is an affine subspace. If M on the other hand is only assumed to be an affine subspace, we may write $M = x_0 + W$ for $x_0 \in M$ and W as above. Now represent W as $\{x \in \mathbb{R}^d \mid Ax = 0\}$ for a suitable $m \times d$ matrix A with $m \le d$ (see Exercise 2.3). Let $M' = \{x \in \mathbb{R}^d \mid Ax = b\}$ with $b = Ax_0$: if $x \in W$, then $A(x_0 + x) = b$ and $M \subseteq M'$. If $z \in M'$, then $z - x_0 \in W$ and the identity $z = x_0 + (z - x_0)$ shows that $M' \subseteq M$ (compare this with the computations in Example 2.1). Therefore $M = M'$.

The last part of the proposition is left as an exercise (Exercise 2.9). $\square$

2.4   Affine independence and the dimension of a subset

Definition 2.7. The *dimension of an affine subspace* $M \subseteq \mathbb{R}^d$ is defined as

$$\dim(M) := \dim W,$$

where W is the subspace $\{u - v \mid u, v \in M\} \subseteq \mathbb{R}^d$ (see Proposition 2.6). The dimension of an arbitrary subset $S \subseteq \mathbb{R}^d$ is defined as

$$\dim S := \dim \operatorname{aff}(S).$$

A finite set $S = \{v_1, \dots, v_m\}$ is called *affinely independent* if $\dim S = m - 1$.

This definition is very intuitive e.g., two points are affinely independent if they are different, three points are affinely independent if they do not lie on the same line etc. A single point has dimension zero, the affine span of two affinely independent points is a line (dimension one), the affine space of three affinely independent points is a plane (dimension two) etc.

If $H \subseteq \mathbb{R}^d$ is an affine subspace of dimension $d - 1$, then $H = V + x_0$, where $x_0 \in H$ and $V = \{x \in \mathbb{R}^d \mid \alpha^t x = 0\}$ for some $\alpha \in \mathbb{R}^d \setminus \{0\}$ i.e., $H = \{x \in \mathbb{R}^d \mid \alpha^t x = \beta\}$ for $\beta = \alpha^t x_0$. Such an affine subspace is called an *affine hyperplane*.

Definition 2.8. If $H = \{x \in \mathbb{R}^n \mid \alpha^t x = \beta\}$ is an *affine hyperplane*, we define

$$H^+ = \{x \in \mathbb{R}^n \mid \alpha^t x \geq \beta\}, \qquad H^{++} = \{x \in \mathbb{R}^n \mid \alpha^t x > \beta\},$$
$$H^- = \{x \in \mathbb{R}^n \mid \alpha^t x \leq \beta\}, \qquad H^{--} = \{x \in \mathbb{R}^n \mid \alpha^t x < \beta\}.$$

The two subsets H^- and H^+ are called *(affine) half spaces*.

Proposition 2.9. *Let* $S = \{v_1, \dots, v_m\} \subseteq \mathbb{R}^d$. *Then* $\operatorname{aff}(S) = v_1 + W$, *where* W *is the subspace spanned by* $v_2 - v_1, \dots, v_m - v_1$. *The following conditions are equivalent.*

(1) *S is affinely independent.*
(2) *$v_2 - v_1, \dots, v_m - v_1$ are linearly independent.*
(3) *The equations*

$$\lambda_1 v_1 + \cdots + \lambda_m v_m = 0$$
$$\lambda_1 + \cdots + \lambda_m = 0$$

imply that $\lambda_1 = \cdots = \lambda_m = 0$.

(4) *The vectors*

$$\begin{pmatrix} v_1 \\ 1 \end{pmatrix}, \ldots, \begin{pmatrix} v_m \\ 1 \end{pmatrix}$$

are linearly independent in $\mathbb{R}^{d+1}$.

Proof. By definition

$$\mathrm{aff}(S) = \{\lambda_1 v_1 + \cdots + \lambda_m v_m \mid \lambda_1 + \cdots + \lambda_m = 1\}$$

and $\{v - v_1 \mid v \in \mathrm{aff}(S)\}$ is the subspace W of Proposition 2.6 for $M = \mathrm{aff}(S)$. Since

$$(\lambda_1 - 1)v_1 + \lambda_2 v_2 + \cdots + \lambda_m v_m = \lambda_2(v_2 - v_1) + \cdots + \lambda_m(v_m - v_1)$$

it follows that W is spanned by $v_2 - v_1, \ldots, v_m - v_1$ as claimed. Therefore (1) is equivalent to (2). To prove (2) $\Rightarrow$ (3) we write

$$\lambda_1 v_1 + \cdots + \lambda_m v_m = \lambda_2(v_2 - v_1) + \cdots + \lambda_m(v_m - v_1)$$

using that $\lambda_1 + \cdots + \lambda_m = 0$. By the linear independence of $v_2 - v_1, \ldots,$ $v_m - v_1$ it follows that $\lambda_2 = \cdots = \lambda_m = 0$ and therefore also $\lambda_1 = 0$. The claim (4) is an exact restatement of (3). Assuming (3) holds, let us finish the proof by showing that (2) holds. Suppose that

$$\mu_1(v_2 - v_1) + \cdots + \mu_{m-1}(v_m - v_1) = 0$$

for $\mu_1, \ldots, \mu_{m-1} \in \mathbb{R}$. We may rewrite this as

$$\lambda v_1 + \mu_1 v_2 + \cdots + \mu_{m-1} v_m$$

with $\lambda = -\mu_1 - \cdots - \mu_{m-1}$. Now (3) implies implies $\mu_1 = \cdots = \mu_{m-1} = 0$ showing that $v_2 - v_1, \ldots, v_m - v_1$ are linearly independent vectors. $\square$

2.5 Exercises

Exercise 2.1. Let $u, v \in \mathbb{R}^d$ with $u \neq v$. Prove that

$$L = \{(1 - t)u + tv \mid t \in \mathbb{R}\}$$

is a line in $\mathbb{R}^d$ containing u and v. Prove also that if M is a line in $\mathbb{R}^d$ such that $u, v \in M$, then $M = L$.

Exercise 2.2. Let $u = (1, 1, 1)$ and $v = (1, 2, 3)$ be vectors in $\mathbb{R}^3$. Show that u and v are linearly independent and find $\alpha \in \mathbb{R}^3$ with

$$W = \{x \in \mathbb{R}^3 \mid \alpha^t x = 0\},$$

where W is the subspace spanned by u and v.

Exercise 2.3. Let $W \subseteq \mathbb{R}^d$ be a subspace and suppose that $v_1, \ldots, v_r$ is a basis of W. Prove that

$$W' = \{u \in \mathbb{R}^d \mid u^t v_1 = \cdots = u^t v_r = 0\} \subseteq \mathbb{R}^d$$

is a subspace. Let $u_1, \ldots, u_s$ be a basis of W' and A the $s \times d$ matrix with these vectors as rows. Show that $s = d - r$ and

$$W = \{x \in \mathbb{R}^d \mid Ax = 0\}.$$

Exercise 2.4. Prove that $\{v \in \mathbb{R}^d \mid Av = b\}$ is an affine subspace of $\mathbb{R}^d$, where A is an $m \times d$ matrix and $b \in \mathbb{R}^m$.

Exercise 2.5. Let M be an affine subspace. Prove that $\{u - v \mid u, v \in M\}$ is a subspace.

Exercise 2.6. Can you have two linearly independent vectors in $\mathbb{R}$? What about two affinely independent vectors?

Exercise 2.7. Decide if $(2, 1)$, $(3, 2)$ and $(5, 5)$ are on the same line in $\mathbb{R}^2$ applying Proposition 2.9.

Exercise 2.8. Let $S = \{(1, 1, 1), (2, 3, 4), (1, 2, 3), (2, 1, 0)\} \subseteq \mathbb{R}^3$. Compute the smallest affine subspace containing S.

Exercise 2.9. Prove that $f(x) = h(x) + b$ is an affine map if $h : \mathbb{R}^d \to \mathbb{R}^n$ is a linear map and $b \in \mathbb{R}^n$. Prove that $h(x) = f(x) - f(0)$ is a linear map if $f : \mathbb{R}^d \to \mathbb{R}^n$ is an affine map.

Exercise 2.10. Prove that you can have no more than $d + 1$ affinely independent vectors in $\mathbb{R}^d$.

Exercise 2.11. Let $v_0, \ldots, v_d$ be affinely independent points in $\mathbb{R}^d$. Prove that

$$f(x) = (\lambda_0, \lambda_1, \ldots, \lambda_d)$$

is a well defined affine map $f : \mathbb{R}^d \to \mathbb{R}^{d+1}$, where

$$x = \lambda_0 v_0 + \cdots + \lambda_d v_d$$

with $\lambda_0 + \cdots + \lambda_d = 1$.

Exercise 2.12. Prove that a non-empty open subset $U \subseteq \mathbb{R}^d$ has dimension $\dim U = d$. Show that a subset $S \subseteq \mathbb{R}^d$ with $\dim S = d$ contains a non-empty open subset.

Chapter 3

Convex subsets

An affine subspace M is the set of solutions to a system of linear equations and contains the line $\{(1 - t)x + ty \,|\, t \in \mathbb{R}\}$ between any two of its points $x, y \in M$. A polyhedron P is the set of solutions to a system of linear inequalities and is only guaranteed to contain the line segment

$$\{(1 - t)x + ty \,|\, 0 \leq t \leq 1\}$$

between any two of its points $x, y \in P$ (see Figure 3.1 and Exercise 3.1).

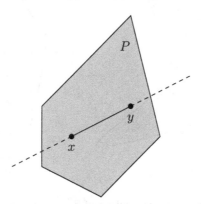

Figure 3.1: A polyhedron $P \subseteq \mathbb{R}^2$ with two points $x, y \in P$, the line segment between x and y and the line through x and y.

In this chapter we will go beyond polyhedra and study subsets of $\mathbb{R}^d$ with the property that they contain the line segment between any two of their points. Subsets with this "nowhere concave" property have appeared up through the history of mathematics at least since the time of Archimedes.[1]

---

[1] Archimedes of Syracuse (287 BC to 212 BC). Greek mathematician.

The first formal definition appeared around 1896 in Minkowski's famous monograph *Geometrie der Zahlen* (see [Kjeldsen (2008)]).

3.1 Basics

The following simple definition (compare with Definition 2.2) is truly fundamental in modern mathematics.

Definition 3.1. A subset $C \subseteq \mathbb{R}^d$ is called *convex* if it contains the line segment between any two of its points:

$$(1 - t)u + tv \in C$$

for every $u, v \in C$ and every $t \in \mathbb{R}$ with $0 \leq t \leq 1$.

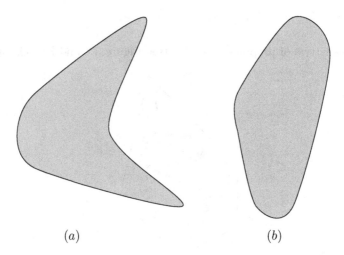

(a)   (b)

Figure 3.2: (a) A non-convex subset of $\mathbb{R}^2$. (b) A convex subset of $\mathbb{R}^2$.

An intersection of convex subsets is a convex subset. Since an affine half space is a convex subset, it follows that polyhedra are convex subsets. The union of convex subsets does not have to be convex.

Minkowski sum, dilation and the polar of a subset

A fundamental operation is setwise addition of convex subsets. The sum $A + B$ of two subsets A and B of $\mathbb{R}^d$ is defined as the subset

$$A + B := \{u + v \mid u \in A, v \in B\} \subseteq \mathbb{R}^d.$$

If A and B are convex subsets, $A + B$ is a convex subset and it is called *the Minkowski sum* of A and B. This rather simple operation on convex subsets turns out to be very important.

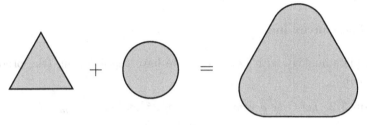

Figure 3.3: Minkowski sum of a triangle and a disc.

Another fundamental operation is *dilation* of a convex subset $C \subseteq \mathbb{R}^d$ by $\lambda \in \mathbb{R}$. This is the convex subset

$$\lambda C := \{\lambda x \mid x \in C\} \subseteq \mathbb{R}^d.$$

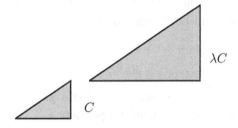

Figure 3.4: Dilation of a convex subset C with $\lambda = 2$. The marked point indicates $(0, 0)$ in $\mathbb{R}^2$.

Let $S \subseteq \mathbb{R}^d$ be any subset. Then

$$S^\circ := \{\alpha \in \mathbb{R}^d \mid \alpha^t x \leq 1, \text{for every } x \in S\} \subseteq \mathbb{R}^d$$

is called the *polar* of S (see Figure 3.5). One can prove that S° is a convex subset of $\mathbb{R}^d$ and that $0 \in S^\circ$.

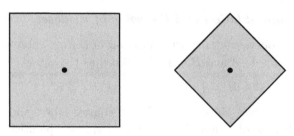

Figure 3.5: A convex subset and its polar. The marked points indicate $(0, 0)$ in $\mathbb{R}^2$.

3.2   The convex hull

In complete analogy with Lemma 2.3 we have the following (the proof is the same).

Lemma 3.2. *Let C be a convex subset of $\mathbb{R}^d$ and $v_1, \dots, v_m \in C$. Then*

$$\lambda_1 v_1 + \cdots + \lambda_m v_m \in C$$

if $\lambda_1, \dots, \lambda_m \geq 0$ and $\lambda_1 + \cdots + \lambda_m = 1$.

A linear combination $\lambda_1 v_1 + \cdots + \lambda_m v_m$ of vectors $v_1, \dots, v_m \in \mathbb{R}^d$ is called convex if $\lambda_1, \dots, \lambda_m \geq 0$ and $\lambda_1 + \cdots + \lambda_m = 1$.

Definition 3.3. The *convex hull of a subset* $S \subseteq \mathbb{R}^d$ is the set of all convex linear combinations of elements from S i.e.,

$$\mathrm{conv}(S) := \{\lambda_1 v_1 + \cdots + \lambda_m v_m \mid m \geq 1,$$
$$v_1, \dots, v_m \in S, \ \lambda_1, \dots, \lambda_m \geq 0 \text{ and } \lambda_1 + \cdots + \lambda_m = 1\}.$$

If S is a finite subset, $\mathrm{conv}(S)$ is called a *polytope*. If $S \subseteq \mathbb{R}^2$ is a finite subset, we call $\mathrm{conv}(S)$ a (convex) *polygon*.

Proposition 3.4. *The convex hull, $\mathrm{conv}(S)$, of a subset $S \subseteq \mathbb{R}^d$ is a convex subset. It is the smallest convex subset containing S.*

The polar set introduced previously has a nice interpretation for the convex hull of a finite set of points.

Proposition 3.5. *Let $C = \mathrm{conv}(\{v_1, \dots, v_m\})$, where $v_1, \dots, v_m \in \mathbb{R}^d$. Then the polar of C is the polyhedron given by*

$$C^\circ = \{\alpha \in \mathbb{R}^d \mid \alpha^t v_1 \leq 1, \dots, \alpha^t v_m \leq 1\}.$$

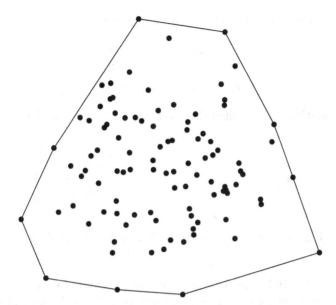

Figure 3.6: The convex hull of 100 random points in $\mathbb{R}^2$. In this example, 90 of the 100 points lie in the convex hull of the 10 "extreme" points. Finding the extreme points is a non-trivial computational task.

Proof. In Exercise 3.9 you are asked to prove this and compute a polar convex hull based on Example 3.6. □

Example 3.6. To get a feeling for convex hulls, it is important to play around with (lots of) examples in the plane. In Figure 3.7 you see a finite subset of four planar points and their convex hull.

In suitable coordinates the four points are

$$\begin{pmatrix} 0 \\ 1 \end{pmatrix}, \quad \begin{pmatrix} 3 \\ 2 \end{pmatrix}, \quad \begin{pmatrix} 2 \\ 1 \end{pmatrix} \quad \text{and} \quad \begin{pmatrix} 4 \\ 0 \end{pmatrix}.$$

You can check that the third point is contained in the convex hull of the other points, since

$$\begin{pmatrix} 2 \\ 1 \end{pmatrix} = \tfrac{3}{7}\begin{pmatrix} 0 \\ 1 \end{pmatrix} + \tfrac{2}{7}\begin{pmatrix} 3 \\ 2 \end{pmatrix} + \tfrac{2}{7}\begin{pmatrix} 4 \\ 0 \end{pmatrix}.$$

You may wonder where the coefficients $\tfrac{3}{7}, \tfrac{2}{7}$ and $\tfrac{2}{7}$ came from. In fact, the coefficients λ_1, λ_2 and λ_3 in

$$\begin{pmatrix} 2 \\ 1 \end{pmatrix} = \lambda_1\begin{pmatrix} 0 \\ 1 \end{pmatrix} + \lambda_2\begin{pmatrix} 3 \\ 2 \end{pmatrix} + \lambda_3\begin{pmatrix} 4 \\ 0 \end{pmatrix}$$

coming from Definition 3.3, must solve the system

$$
\begin{aligned}
3\lambda_2 + 4\lambda_3 &= 2 \\
\lambda_1 + 2\lambda_2 &= 1 \\
\lambda_1 + \lambda_2 + \lambda_3 &= 1
\end{aligned}
$$

of linear equations. This makes them very explicit and computable.

Figure 3.7: Four points (left). The convex hull (right).

One of the points in Figure 3.7 can be omitted without altering the convex hull. This point seems not to be a "vertex" or a "corner". We will give a quite general definition that captures the "corners" or "vertices" of a convex subset.

3.3   Faces of convex subsets

A convex subset C is structured around its faces, which are special convex subsets of C. Informally the zero-dimensional faces are the vertices of C, the one-dimensional faces the "extreme" line segments of C etc. The vertices (called extreme points) are the points not located in the "interior" of a line segment in C. Only one of the marked points in the triangle to the right in Figure 3.7 fails to be extreme. The extreme points are the vertices and the one-dimensional faces the edges of the triangle. We will prove later (see Theorem 6.12) that a compact convex subset always is the convex hull of its extreme points. The precise definition of a face in a convex subset is as follows.

Definition 3.7. Let $C \subseteq \mathbb{R}^d$ be a convex subset. A subset $F \subseteq C$ is called a *face* of C if F is convex and for every $x, y \in C$ and $0 < \lambda < 1$,

$$
(1 - \lambda)x + \lambda y \in F
$$

implies that $x, y \in F$.

An important type of faces of a convex subset is given by following result.

Lemma 3.8. *Let $C \subseteq \mathbb{R}^d$ be a convex subset and $\alpha \in \mathbb{R}^d$. Then*

$$F = \{z \in C \,|\, \alpha^t z \leq \alpha^t x, \text{ for every } x \in C\} \tag{3.1}$$

is a face of C.

Proof. If $z_1, z_2 \in F$, then $\alpha^t z_1 \leq \alpha^t z_2$ and $\alpha^t z_2 \leq \alpha^t z_1$. Therefore $\alpha^t z_1 = \alpha^t z_2$ and $F = \{x \in C \,|\, \alpha^t x = \beta\}$ with $\beta = \alpha^t z_1$, which shows that F is a convex subset as the intersection of C with an affine hyperplane.

Suppose that $x, y \in C$ and $z := (1 - \lambda)x + \lambda y \in F$ for $0 < \lambda < 1$ i.e., $\alpha^t z \leq \alpha^t v$ for every $v \in C$, in particular $\alpha^t z \leq \alpha^t x$ and $\alpha^t z \leq \alpha^t y$. Let us assume that $\alpha^t x \leq \alpha^t y$. Then

$$\alpha^t x = (1 - \lambda)\alpha^t x + \lambda \alpha^t x \leq (1 - \lambda)\alpha^t x + \lambda \alpha^t y = \alpha^t z.$$

This implies $x \in F$. Since we must have $\alpha^t y = \alpha^t x$, it also follows that $y \in F$. $\square$

A face $F \subseteq C$ of a convex subset $C \subseteq \mathbb{R}^d$ is called *exposed* if it is given as in (3.1) for some $\alpha \in \mathbb{R}^d$. A face of a convex subset does not have to be exposed (see Exercise 3.11). A zero-dimensional face (a point) is called an *extreme point*. A point $z \in C$ is extreme if for every $x, y \in C$

$$z \in \text{conv}(\{x, y\})$$

if and only if $z = x$ or $z = y$. The set of extreme points in C is denoted $\text{ext}(C)$.

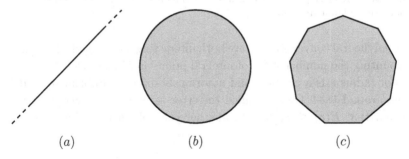

(a) $\qquad\qquad$ (b) $\qquad\qquad$ (c)

Figure 3.8: (a) A line in the plane does not have extreme points. (b) A disc has infinitely many extreme points but no one-dimensional faces. (c) A polygon with nine extreme points and nine one-dimensional faces.

The notion of an extreme point identifies the non-redundant points in a convex hull (see Figure 3.6). Formally, one can show (see Exercise 3.13) by induction that if z is an extreme point of $\text{conv}(\{x_1, \ldots, x_n\})$, then

$$z = x_1 \quad \text{or} \quad z = x_2 \quad \text{or} \quad \ldots \quad \text{or} \quad z = x_n.$$

Interlude: Integral points in convex subsets

This section is meant to be a leisurely break from the formal theory and a pointer to some of the rich and beautiful mathematics that surrounds convexity.

A convex subset does not have to be a polyhedron. An example of non-polyhedral convex subset is a planar disc (see Figure 3.9).

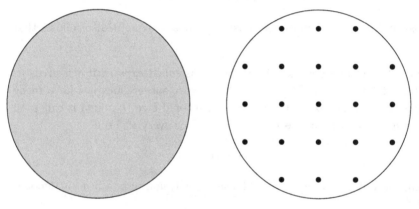

Figure 3.9: (left) A planar disc is not a polyhedron. (right) The integral points of a planar disc.

Non-polyhedral convex subsets are both interesting and difficult. For one thing, counting the number $N(r)$ of integral points inside a circle (see Figure 3.9) of radius r is a classical and notoriously difficult problem studied by Gauss around 1834. Gauss studied the error term $E(r) = |N(r) - \pi r^2|$ and proved that $E(r) \leq 2\sqrt{2}\pi r$. In another example, a four-dimensional ball

$$B(N) = \{(x_1, x_2, x_3, x_4) \in \mathbb{R}^4 \mid x_1^2 + x_2^2 + x_3^2 + x_4^2 \leq N\} \subseteq \mathbb{R}^4$$

is a convex subset. There is a remarkable connection between the prime factorization of $N \in \mathbb{N}$ and counting integral points in $B(N)$. Notice that

$$|B(N) \cap \mathbb{Z}^4| - |B(N-1) \cap \mathbb{Z}^4|$$
$$= \left|\{(x_1, x_2, x_3, x_4) \in \mathbb{Z}^4 \mid x_1^2 + x_2^2 + x_3^2 + x_4^2 = N\}\right|.$$

An amazing result [Jacobi (1829)] due to Jacobi[2] about the number of ways of writing an integer as a sum of four squares, says that

$$\left|\{(x_1, x_2, x_3, x_4) \in \mathbb{Z}^4 \mid x_1^2 + x_2^2 + x_3^2 + x_4^2 = N\}\right| = 8 \sum_{d|N, 4\nmid d} d. \qquad (3.2)$$

As a simple check of this remarkable formula you can verify that 4 is a sum of four squares in the following $8 + 16$ different ways:

$$(\pm 2, 0, 0, 0), \quad (0, \pm 2, 0, 0), \quad (0, 0, \pm 2, 0), \quad (0, 0, 0, \pm 2)$$
$$(\pm 1, \pm 1, \pm 1, \pm 1).$$

If you consider a number $N = pq$, which is the product of two odd prime numbers p and q, you can use (3.2) to find $p+q$ by counting integral points in $B(N)$. But if you know $N = pq$ and $p+q$, then you can retrieve p and q, since

$$(x-p)(x-q) = x^2 - (p+q)x + N.$$

This may sound esoteric, but in essence, it shows that if you can count integral points in the four-dimensional ball effectively, then you can crack the worldwide encryption system RSA (RSA is based on the computational hardness of prime factorization – we are talking 500 digit numbers or more here).

Counting integral points in polyhedra is difficult but much better understood. For example if P is a convex polygon with $\dim(P) = 2$ and integral vertices, then the number of integral points in P is given by the formula of Pick[3] from 1899:

$$|P \cap \mathbb{Z}^2| = \text{Area}(P) + \tfrac{1}{2}\,\text{B}(P) + 1,$$

where $\text{B}(P)$ is the number of integral points on the boundary of P. You can easily check this with a few examples. Consider for example the convex polygon P in Figure 3.10.

---

[2]Carl Gustav Jacobi (1804–1851). German mathematician.
[3]Georg Alexander Pick (1859–1942). Austrian mathematician.

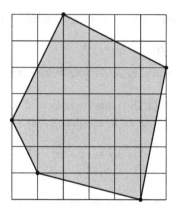

Figure 3.10: Integral points in a convex polygon with integral vertices.

By subdivision into triangles it follows that $\text{Area}(P) = \frac{55}{2}$. Also, by an easy count we get $B(P) = 7$. Therefore the formula of Pick shows that

$$|P \cap \mathbb{Z}^2| = \frac{55}{2} + \frac{1}{2} \cdot 7 + 1 = 32.$$

The polygon contains 32 integral points. This can be verified by an explicit count from Figure 3.10.

For a very nice account of counting integral points in polyhedra and beyond look up [de Loera (2005)].

An exceedingly important class of convex subsets is now introduced and the connection to the convex hull is described.

3.4   Convex cones

A *cone* in $\mathbb{R}^d$ is a subset K such that $\lambda x \in K$ for every $x \in K$ and every $\lambda \geq 0$. A *pointed cone* is a cone not containing a line.

Proposition 3.9. *A non-empty polyhedron $P = \{x \in \mathbb{R}^d \,|\, Ax \leq b\}$ is a cone if and only if $P = \{x \in \mathbb{R}^d \,|\, Ax \leq 0\}$. P is a pointed cone if and only if $\text{rk}\, A = d$.*

Proof. If $P = \{x \in \mathbb{R}^d \,|\, Ax \leq 0\}$, then P is a cone, since $A(\lambda x) = \lambda(Ax) \leq 0$ if $\lambda \geq 0$ and $Ax \leq 0$. If $P = \{x \in \mathbb{R}^d \,|\, Ax \leq b\}$ is a cone, then $b \geq 0$, since $0 \in P$. Therefore $\{x \in \mathbb{R}^d \,|\, Ax \leq 0\} \subseteq P$. If $x \in P$ and $Ax \not\leq 0$, then $a_j^t x > 0$ for some row vector a_j of A and $a_j^t(\lambda x) = \lambda(a_j^t x) \leq b_j$ for every $\lambda \geq 0$, since P is a cone. This contradicts that $a_j^t z$ is bounded above by b_j for $z \in P$. Therefore $P = \{x \in \mathbb{R}^d \mid Ax \leq 0\}$.

If $\operatorname{rk} A < d$, there exists a nonzero $\alpha \in \mathbb{R}^d$ with $A\alpha = 0$. This shows that P contains the line $\{t\alpha \mid t \in \mathbb{R}\}$. On the other hand, suppose that P contains the line $L = \{\beta + t\alpha \mid t \in \mathbb{R}\}$ with $\alpha \in \mathbb{R}^d \setminus \{0\}$. If $a_j^t \alpha \neq 0$ for some j, then we may find $t \in \mathbb{R}$ such that $a_j^t \beta + t a_j^t \alpha > 0$ contradicting that $L \subseteq P$. Therefore $A\alpha = 0$ and we must have $\operatorname{rk} A < d$. $\qquad\square$

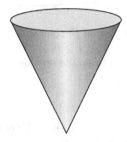

Figure 3.11: Part of the Lorentz cone in $\mathbb{R}^3$.

A polyhedral cone is a convex cone. An example of a non-polyhedral convex cone is the *Lorentz cone* (see Figure 3.11) given by

$$C = \{(x, y, z) \in \mathbb{R}^3 \mid z \geq 0,\ x^2 + y^2 \leq z^2\}.$$

It is not too hard to prove that C is a cone. Proving that C is a convex subset is more of a challenge (Exercise 3.17).

The recession cone

To every convex subset $C \subseteq \mathbb{R}^n$ we associate a convex cone denoted $\operatorname{rec}(C)$ called the *recession cone* of C. Informally $\operatorname{rec}(C)$ is the set of directions in which C recedes[4] (see [Rockafellar (1970)], Section 8). The precise definition is

$$\operatorname{rec}(C) = \{d \in \mathbb{R}^n \mid x + d \in C,\ \text{for every } x \in C\}. \qquad (3.3)$$

It is left as an exercise (Exercise 3.18) to prove that $\operatorname{rec}(C)$ is a convex cone. If C is a convex cone, then $\operatorname{rec}(C) = C$.

---

[4]Go to or toward a more distant point.

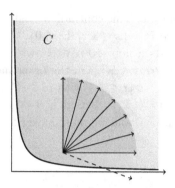

Figure 3.12: The recession cone of the convex subset $C = \{(x,y) \mid xy \geq 1,$
$x > 0\} \subseteq \mathbb{R}^2$ is $\text{rec}(C) = \{(x,y) \mid x \geq 0, y \geq 0\}$. The dashed arrow is not
in $\text{rec}(C)$. Its direction leads to points outside C.

Finitely generated cones

A linear combination $\lambda_1 v_1 + \cdots + \lambda_m v_m$ of vectors $v_1, \ldots, v_m \in \mathbb{R}^d$ is
called *conic* if $\lambda_1, \ldots, \lambda_m \geq 0$. In complete analogy with the convex hull we
introduce the following.

Definition 3.10. The cone generated by a subset $S \subseteq \mathbb{R}^d$ is the set of all
conic linear combinations of elements from S i.e.,

$$\text{cone}(S) := \{\lambda_1 v_1 + \cdots + \lambda_m v_m \mid m \geq 1,$$
$$v_1, \ldots, v_m \in S, \lambda_1, \ldots, \lambda_m \geq 0\}.$$

A convex cone C is called *finitely generated* if $C = \text{cone}(S)$ for a finite set
$S \subseteq \mathbb{R}^d$.

Convex cones are closely related to the convex hull. Here we repeat the idea
illustrated in Example 3.6 in the context of finitely generated cones.

Example 3.11. A triangle T is the convex hull of three affinely independent points

$$(x_1, y_1), \quad (x_2, y_2), \quad (x_3, y_3)$$

in the plane and $(x, y) \in T$ if and only if

$$\begin{pmatrix} x \\ y \\ 1 \end{pmatrix} \in \text{cone}\left(\left\{\begin{pmatrix} x_1 \\ y_1 \\ 1 \end{pmatrix}, \begin{pmatrix} x_2 \\ y_2 \\ 1 \end{pmatrix}, \begin{pmatrix} x_3 \\ y_3 \\ 1 \end{pmatrix}\right\}\right). \tag{3.4}$$

Testing (3.4) amounts to solving the system

$$\begin{pmatrix} x_1 & x_2 & x_3 \\ y_1 & y_2 & y_3 \\ 1 & 1 & 1 \end{pmatrix} \begin{pmatrix} \lambda_1 \\ \lambda_2 \\ \lambda_3 \end{pmatrix} = \begin{pmatrix} x \\ y \\ 1 \end{pmatrix} \tag{3.5}$$

of linear equations. So $(x, y) \in T$ if and only if the unique solution to (3.5) has $\lambda_1 \geq 0$, $\lambda_2 \geq 0$ and $\lambda_3 \geq 0$. Let us experiment with a few concrete numbers. You can plot the points

$$\begin{pmatrix} 0 \\ 0 \end{pmatrix}, \begin{pmatrix} 2 \\ 1 \end{pmatrix} \quad \text{and} \quad \begin{pmatrix} 5 \\ 3 \end{pmatrix} \in \mathbb{R}^2$$

and realize that their convex hull, T, is a very thin triangle. From a drawing it can be difficult to decide if a given point is inside the triangle. Here the 3×3 matrix from (3.5) helps. In this case

$$A = \begin{pmatrix} 0 & 2 & 5 \\ 0 & 1 & 3 \\ 1 & 1 & 1 \end{pmatrix}$$

and

$$A^{-1} = \begin{pmatrix} -2 & 3 & 1 \\ 3 & -5 & 0 \\ -1 & 2 & 0 \end{pmatrix}.$$

Let us check if $v_1 = (1, \frac{11}{20}) \in T$ and $v_2 = (4, 2) \in T$. In the case of v_1,

$$A^{-1} \begin{pmatrix} 1 \\ 11/20 \\ 1 \end{pmatrix} = \begin{pmatrix} 13/20 \\ 1/4 \\ 1/10 \end{pmatrix}.$$

This is the solution $(\lambda_1, \lambda_2, \lambda_3)$ to the linear equations in (3.5). In this case, the solution satisfies $\lambda_1 \geq 0$, $\lambda_2 \geq 0$ and $\lambda_3 \geq 0$. Therefore $v_1 \in T$. As an added bonus you also see that

$$\begin{pmatrix} 1 \\ \frac{11}{20} \end{pmatrix} = \frac{13}{20} \begin{pmatrix} 0 \\ 0 \end{pmatrix} + \frac{1}{4} \begin{pmatrix} 2 \\ 1 \end{pmatrix} + \frac{1}{10} \begin{pmatrix} 5 \\ 3 \end{pmatrix}.$$

For v_2 we get

$$A^{-1} \begin{pmatrix} 4 \\ 2 \\ 1 \end{pmatrix} = \begin{pmatrix} -1 \\ 2 \\ 0 \end{pmatrix}.$$

Here the first coordinate of the solution to (3.5) is negative and therefore $v_2 \notin T$.

In general, the *polar of a subset* $S \subseteq \mathbb{R}^d$ is defined as

$$S^\circ = \{\alpha \in \mathbb{R}^d \,|\, \alpha^t x \leq 1 \text{ for every } x \in S\}. \tag{3.6}$$

If S happens to be a cone K we can replace the upper bound of 1 in (3.6) by 0, because if $\alpha^t x > 0$ for some $x \in K$, we would have $\alpha^t(\lambda x) = \lambda(\alpha^t x) > 1$ for a sufficiently big $\lambda > 0$. This contradicts that $\alpha \in K^\circ$, since $\lambda x \in K$. Therefore

$$K^\circ = \{\alpha \in \mathbb{R}^d \,|\, \alpha^t x \leq 0 \text{ for every } x \in K\}.$$

The *polar cone* K° is a cone and therefore a convex cone. In complete analogy with Proposition 3.5 we have the following result.

Proposition 3.12. *If $C \subseteq \mathbb{R}^d$ is a finitely generated cone, then C° is a polyhedral cone.*

Proof. Suppose that $C = \text{cone}(S)$, where $S = \{v_1, \ldots, v_m\} \subseteq \mathbb{R}^d$. Then $\alpha^t x \leq 0$ for every $x = \lambda_1 v_1 + \cdots + \lambda_m v_m$, where $\lambda_i \geq 0$ for $i = 1, \ldots, m$ if and only if $\alpha^t v_i \leq 0$ for $i = 1, \ldots, m$. Let A be the $m \times d$ matrix with $v_1, \ldots, v_m$ as its rows. Then it follows that

$$C^\circ = \{\alpha \in \mathbb{R}^n \,|\, A\alpha \leq 0\}. \qquad \square$$

Polyhedral cones are finitely generated and finitely generated cones are polyhedral. These two deeper results due to Minkowski and Weyl are the focus of the next chapter. In purely algebraic terms, the finite generation of a polyhedral cone says that there exists finitely many solutions $v_1, \ldots, v_N \in \mathbb{R}^n$ to a system of linear inequalities

$$\begin{aligned} a_{11}x_1 + \cdots + a_{n1}x_n &\leq 0 \\ &\;\;\vdots \\ a_{1m}x_1 + \cdots + a_{nm}x_n &\leq 0, \end{aligned} \tag{3.7}$$

such that every solution to (3.7) is a conic linear combination of $v_1, \ldots, v_N$. In the setting of linear algebra you know that every solution to the homogeneous linear system corresponding (3.7) is a linear combination of $\leq n$ solutions. For (3.7) even the existence of finitely many (conic) generating solutions is a non-trivial statement.

Example 3.13. In Figure 3.13 we have sketched a finitely generated cone C and its polar cone C°. If you look closer, you will see that

$$C = \text{cone}\left(\left\{\begin{pmatrix}2\\1\end{pmatrix}, \begin{pmatrix}1\\2\end{pmatrix}\right\}\right) \quad \text{and} \quad C^\circ = \text{cone}\left(\left\{\begin{pmatrix}1\\-2\end{pmatrix}, \begin{pmatrix}-2\\1\end{pmatrix}\right\}\right).$$

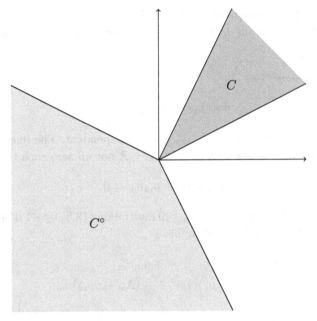

Figure 3.13: A finitely generated cone C and its polar cone C°.

Notice also that C° encodes the fact that C is the intersection of the two half planes

$$\left\{ \begin{pmatrix} x \\ y \end{pmatrix} \in \mathbb{R}^2 \left| \begin{pmatrix} 1 \\ -2 \end{pmatrix}^t \begin{pmatrix} x \\ y \end{pmatrix} \leq 0 \right. \right\} \quad \text{and} \quad \left\{ \begin{pmatrix} x \\ y \end{pmatrix} \in \mathbb{R}^2 \left| \begin{pmatrix} -2 \\ 1 \end{pmatrix}^t \begin{pmatrix} x \\ y \end{pmatrix} \leq 0 \right. \right\}.$$

3.5 Carathéodory's theorem

In Example 3.11 you saw how to check if a point is in the convex hull of three points in the plane. This led to three linear equations in three unknowns. What if you had more than three points? In §3.6 it is shown how to reduce such a computation to a finite number of the simple computations carried out in Example 3.11. The key point is that a finitely generated cone is the union of finitely many simplicial subcones. A convex cone is called *simplicial* if it is generated by finitely many linearly independent vectors. The following result was first proved by Carathéodory.[5]

---

[5]Constantin Carathéodory (1873–1950). Greek mathematician.

Theorem 3.14 (Carathéodory). *Let* $V = \{v_1, \ldots, v_m\} \subseteq \mathbb{R}^d$. *If* $v \in$ cone(V), *then* v *belongs to the cone generated by a linearly independent subset of* V.

Proof. Suppose that

$$v = \lambda_1 v_1 + \cdots + \lambda_m v_m$$

with $\lambda_1, \ldots, \lambda_m > 0$ and $v_1, \ldots, v_m$ linearly dependent. The linear dependence means that there exists $\mu_1, \ldots, \mu_m \in \mathbb{R}$ not all zero such that

$$\mu_1 v_1 + \cdots + \mu_m v_m = 0. \tag{3.8}$$

We may assume that at least one $\mu_i > 0$ multiplying (3.8) by -1 if necessary. Introducing the variable $\theta \in \mathbb{R}$ we have

$$\begin{aligned} v &= v - \theta(\mu_1 v_1 + \cdots + \mu_m v_m) \\ &= (\lambda_1 - \theta\mu_1)v_1 + \cdots + (\lambda_m - \theta\mu_m)v_m. \end{aligned} \tag{3.9}$$

Let

$$\begin{aligned} \theta^* &= \max\{\theta \geq 0 \,|\, \lambda_i - \theta\mu_i \geq 0, \text{ for every } i = 1, \ldots, m\} \\ &= \min\left\{\frac{\lambda_i}{\mu_i} \,\middle|\, \mu_i > 0, \ i = 1, \ldots, m\right\}. \end{aligned}$$

When you substitute θ^* for θ in (3.9), you discover that v lies in the subcone generated by a proper subset of V. Basically we are varying θ in (3.9) ensuring non-negative coefficients for $v_1, \ldots, v_m$ until "the first time" we reach a zero coefficient in front of some v_j. This (or these) v_j is (are) deleted from the generating set. Repeating this process we will eventually end up with a linearly independent subset of vectors from V. □

Corollary 3.15. *Let* $V = \{v_1, \ldots, v_m\} \subseteq \mathbb{R}^d$. *If* $v \in$ conv(V), *then* v *belongs to the convex hull of an affinely independent subset of* V.

Proof. If $v \in$ conv$(\{v_1, \ldots, v_m\})$, then

$$\begin{pmatrix} v \\ 1 \end{pmatrix} \in \text{cone}\left(\left\{\begin{pmatrix} v_1 \\ 1 \end{pmatrix}, \ldots, \begin{pmatrix} v_m \\ 1 \end{pmatrix}\right\}\right).$$

Now use Theorem 3.14 to conclude that

$$\begin{pmatrix} v \\ 1 \end{pmatrix} \in \text{cone}\left(\left\{\begin{pmatrix} u_1 \\ 1 \end{pmatrix}, \ldots, \begin{pmatrix} u_k \\ 1 \end{pmatrix}\right\}\right),$$

where

$$\left\{\binom{u_1}{1}, \ldots, \binom{u_k}{1}\right\} \subseteq \left\{\binom{v_1}{1}, \ldots, \binom{v_m}{1}\right\}$$

is a linearly independent subset. Therefore $v \in \text{conv}(\{u_1, \ldots, u_k\})$. But

$$\binom{u_1}{1}, \ldots, \binom{u_k}{1}$$

are linearly independent if and only if $u_1, \ldots, u_k$ are affinely independent by Proposition 2.9. $\qquad \square$

A consequence of Corollary 3.15 is that a point in the convex hull of more than three planar points, belongs to the convex hull of at most three of the points (see Figure 3.14).

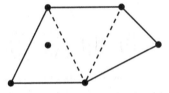

Figure 3.14: A point in the convex hull of some given planar points is in the convex hull of at most three of these points.

The affine equivalent of a simplicial cone is called a *simplex*. More precisely, a convex subset is called a *d-simplex* if it is the convex hull of $d + 1$ affinely independent points (see Figure 3.15). In these terms, Corollary 3.15 states that the convex hull of finitely many points is a union of finitely many simplices.

Figure 3.15: Picture of 0-simplex (point), 1-simplex (line segment), 2-simplex (triangle), 3-simplex (tetrahedron).

Example 3.16. The proofs of Theorem 3.14 and Corollary 3.15 may appear quite abstract until you study a few concrete examples. Here is a typical example illustrating the linear algebra that goes into their proofs. Consider the convex hull

$$C = \text{conv}\left(\left\{\binom{1}{0}, \binom{0}{1}, \binom{1}{2}, \binom{4}{0}\right\}\right).$$

The identity

$$\binom{2}{1} = \tfrac{1}{11}\binom{1}{0} + \tfrac{1}{11}\binom{0}{1} + \tfrac{5}{11}\binom{1}{2} + \tfrac{4}{11}\binom{4}{0}$$

shows that

$$v = \binom{2}{1} \in C.$$

Corollary 3.15 says however that v is in the convex hull of at most three of these four vectors (why?). Let us emulate the proof of Corollary 3.15 in a concrete computation. The first step is to find a linear dependence between the four vectors

$$\begin{pmatrix}1\\0\\1\end{pmatrix}, \quad \begin{pmatrix}0\\1\\1\end{pmatrix}, \quad \begin{pmatrix}1\\2\\1\end{pmatrix} \quad \text{and} \quad \begin{pmatrix}4\\0\\1\end{pmatrix} \tag{3.10}$$

in $\mathbb{R}^3$ (compare this with the beginning of the proof of Corollary 3.15). Here is a linear dependence:

$$5\begin{pmatrix}1\\0\\1\end{pmatrix} - 6\begin{pmatrix}0\\1\\1\end{pmatrix} + 3\begin{pmatrix}1\\2\\1\end{pmatrix} - 2\begin{pmatrix}4\\0\\1\end{pmatrix} = \begin{pmatrix}0\\0\\0\end{pmatrix}. \tag{3.11}$$

Just to make sure you know how to translate linear dependence into the more mundane setting of linear equations, let us show (in painstaking detail) how (3.11) was found! A linear dependence for the vectors in (3.10) can be found as a non-zero solution to the equations

$$\begin{array}{rcrcrcrcl}
\lambda_1 & & & + & \lambda_3 & + & 4\lambda_4 & = & 0 \\
& & \lambda_2 & + & 2\lambda_3 & & & = & 0 \\
\lambda_1 & + & \lambda_2 & + & \lambda_3 & + & \lambda_4 & = & 0
\end{array} \tag{3.12}$$

in $\lambda_1, \lambda_2, \lambda_3$ and λ_4. Such a non-zero solution always exists (see Theorem B.2). By subtracting the first equation from the third (a step in Gaussian elimination), you get

$$\begin{array}{rcrcrcl}
\lambda_1 & + & \lambda_3 & + & 4\lambda_4 & = & 0 \\
\lambda_2 & + & 2\lambda_3 & & & = & 0 \\
\lambda_2 & & & - & 3\lambda_4 & = & 0.
\end{array}$$

From this system of equations you can glean the non-zero solution

$$\lambda_2 = -6, \quad \lambda_4 = -2, \quad \lambda_3 = 3, \quad \text{and} \quad \lambda_1 = 5,$$

which is the linear dependence in (3.11). In this particular case (3.9) becomes

$$\binom{2}{1} = (\tfrac{1}{11} - \theta 5)\begin{pmatrix}1\\0\\1\end{pmatrix} + (\tfrac{1}{11} + \theta 6)\begin{pmatrix}0\\1\\1\end{pmatrix} + (\tfrac{5}{11} - \theta 3)\begin{pmatrix}1\\2\\1\end{pmatrix} + (\tfrac{4}{11} + \theta 2)\begin{pmatrix}4\\0\\1\end{pmatrix}.$$

How big is θ allowed to be, when

$$\tfrac{1}{11} - \theta\,5 \geq 0$$
$$\tfrac{5}{11} - \theta\,3 \geq 0$$

must hold? Solving the inequalities for θ we get $\tfrac{1}{55} \geq \theta$ and $\tfrac{5}{33} \geq \theta$. To ensure that both inequalities are satisfied, we must have $\tfrac{1}{55} \geq \theta$. The magic appears when we pick $\theta = \tfrac{1}{55}$. Then

$$\begin{pmatrix}2\\1\\1\end{pmatrix} = (\tfrac{1}{11} - \tfrac{1}{55}5)\begin{pmatrix}1\\0\\1\end{pmatrix} + (\tfrac{1}{11} + \tfrac{1}{55}6)\begin{pmatrix}0\\1\\1\end{pmatrix} + (\tfrac{5}{11} - \tfrac{1}{55}3)\begin{pmatrix}1\\2\\1\end{pmatrix} + (\tfrac{4}{11} + \tfrac{1}{55}2)\begin{pmatrix}4\\0\\1\end{pmatrix}$$

$$= \tfrac{11}{55}\begin{pmatrix}0\\1\\1\end{pmatrix} + \tfrac{22}{55}\begin{pmatrix}1\\2\\1\end{pmatrix} + \tfrac{22}{55}\begin{pmatrix}4\\0\\1\end{pmatrix}.$$

This implies

$$\begin{pmatrix}2\\1\end{pmatrix} = \tfrac{11}{55}\begin{pmatrix}0\\1\end{pmatrix} + \tfrac{22}{55}\begin{pmatrix}1\\2\end{pmatrix} + \tfrac{22}{55}\begin{pmatrix}4\\0\end{pmatrix}.$$

Therefore

$$\begin{pmatrix}2\\1\end{pmatrix} \in \mathrm{conv}\left(\begin{pmatrix}0\\1\end{pmatrix}, \begin{pmatrix}1\\2\end{pmatrix}, \begin{pmatrix}4\\0\end{pmatrix}\right).$$

3.6 The convex hull, simplicial subsets and Bland's rule

How do we decide if a point is in the convex hull of finitely many given points in $\mathbb{R}^d$? You have seen in the proof of Corollary 3.15 and in Example 3.16 how this can be reduced to deciding if a vector is a conic linear combination of finitely many given vectors. In principle, Theorem 3.14 tells us how to check this. But if V in the statement of Theorem 3.14 is a generating set for $\mathbb{R}^d$ with m elements, we need to check the linearly independent subsets of V with d elements. This seems like an insurmountable problem. If for example $m = 30$ and $d = 10$, there can be up to $30\,045\,015$ of the latter.

We will present a rather surprising algorithm for traversing these subsets coming from §7.1 in the classic monograph [Schrijver (1986)]. The surprise is that the algorithm does not get stuck — it manages to traverse the linearly independent subsets until it finds that the vector is in the cone or that this is certifiably false. The method is inspired by Bland's rule in the simplex algorithm (see §5.4).

Let $V = \{v_1, \ldots, v_m\} \subseteq \mathbb{R}^d$ be a generating set and $x \in \mathbb{R}^d$. We call a subset $J \subseteq \{1, 2, \ldots, m\}$ of size d simplicial if $|J| = d$ and $\{v_j \mid j \in J\}$ is linearly independent.

(1) For a simplicial subset $J = \{i_1, \ldots, i_d\} \subseteq \{1, 2, \ldots, m\}$, there exists uniquely defined $\lambda_{i_j} \in \mathbb{R}$ for $j = 1, \ldots, d$, such that

$$x = \lambda_{i_1} v_{i_1} + \cdots + \lambda_{i_d} v_{i_d}.$$

(2) If $\lambda_{i_j} \geq 0$ for every $j = 1, \ldots, d$, we have verified that $v \in \mathrm{cone}(V)$. If not, pick the *smallest* index $k \in \{i_1, \ldots, i_d\}$ with $\lambda_k < 0$. Let $c \in \mathbb{R}^d$ be the unique vector with the property that

$$c^t v_k = 1$$
$$c^t v_j = 0, \qquad \text{if } j \in \{i_1, \ldots, i_d\} \setminus \{k\}.$$

This vector is the k-th row in A^{-1}, where A is the matrix with column vectors $v_{i_1}, \ldots, v_{i_d}$. Notice that $c^t x = \lambda_k < 0$.

(3) If $c^t v_j \geq 0$ for every $j \notin J$, then $c^t v \geq 0$ for every $v \in \mathrm{cone}(V)$. Since $c^t x < 0$, this shows that $x \in \mathrm{cone}(V)$ is impossible.

(4) If not, pick the *smallest* index $j \in \{1, \ldots, m\}$ with $c^t v_j < 0$ and repeat the procedure from (1) with the simplicial subset

$$J := (J \setminus \{k\}) \cup \{j\}.$$

Here we will say that j enters and k exits the simplicial subset.

Example 3.17. As an illustration (see Figure 3.16) let us check if

$$\begin{pmatrix} 1 \\ 1 \end{pmatrix} \in \mathrm{conv}\left(\left\{ \begin{pmatrix} 1 \\ 0 \end{pmatrix}, \begin{pmatrix} 0 \\ 1 \end{pmatrix}, \begin{pmatrix} 2 \\ 2 \end{pmatrix}, \begin{pmatrix} 5 \\ 1 \end{pmatrix}, \begin{pmatrix} 4 \\ 0 \end{pmatrix} \right\} \right).$$

Rewriting this as a conic problem we need to check if $x \in \mathrm{cone}(V)$, where

$$V = \left\{ \begin{pmatrix} 1 \\ 0 \\ 1 \end{pmatrix}, \begin{pmatrix} 0 \\ 1 \\ 1 \end{pmatrix}, \begin{pmatrix} 2 \\ 2 \\ 1 \end{pmatrix}, \begin{pmatrix} 5 \\ 1 \\ 1 \end{pmatrix}, \begin{pmatrix} 4 \\ 0 \\ 1 \end{pmatrix} \right\} \quad \text{and} \quad x = \begin{pmatrix} 1 \\ 1 \\ 1 \end{pmatrix}.$$

We initiate the algorithm above with the simplicial subset $J = \{3, 4, 5\}$. Here

$$x = v_3 - v_4 + v_5$$

with $k = 4$ and $c = (\frac{1}{2}, \frac{1}{2}, -2)$. Since $c^t v_1 = -\frac{3}{2}$ and $c^t v_2 = -2$ we get $j = 1$ and repeat the iteration with $J = \{3, 1, 5\}$. Here

$$x = \tfrac{1}{2} v_3 + \tfrac{2}{3} v_1 - \tfrac{1}{6} v_5$$

with $k = 5$ and $c = (\frac{1}{3}, -\frac{1}{6}, -\frac{1}{3})$. Since $c^t v_2 = -\frac{1}{2}$ and $c^t v_4 = \frac{7}{6}$ we get $j = 2$. Therefore the iteration is repeated with $J = \{3, 1, 2\}$. Here

$$x = \tfrac{1}{3} v_3 + \tfrac{1}{3} v_1 + \tfrac{1}{3} v_2$$

showing that $x \in \mathrm{cone}(V)$.

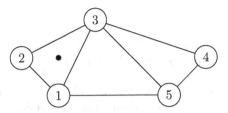

Figure 3.16: Geometric illustration of the steps $\{3,4,5\} \rightarrow \{1,3,5\} \rightarrow \{1,2,3\}$ in detecting that the marked point is in the convex hull of the points in Example 3.17.

Non-cycling

We will now prove that the procedure outlined in the above steps (1)–(4) terminates. Let J_t denote the simplicial subset in step t of the algorithm. If the algorithm does not terminate it must enter into a cycle: $J_k = J_l$ for some $k < l$.

Let $r \in \{1, 2, \ldots, m\}$ be the largest index to exit the simplicial subset in a step $p \in \{k, k+1, \ldots, l-1\}$. Since $J_k = J_l$, there must also exist a step $q \in \{k, k+1, \ldots, l-1\}$ for which r enters the simplicial subset. Suppose that $J_p = \{i_1, \ldots, i_d\}$. Then

$$x = \lambda_{i_1} v_{i_1} + \cdots + \lambda_{i_d} v_{i_d} \tag{3.13}$$

with $\lambda_{i_j} \geq 0$ for $i_j < r$ and $\lambda_r < 0$.

Let c denote the vector of (2) in step q, where r enters. Then

$$c^t x < 0. \tag{3.14}$$

Since r enters the simplicial subset, we must have $c^t v_{i_j} \geq 0$ for $i_j < r$ and $c^t v_r < 0$ by (4). We already know that $\lambda_r < 0$ and $\lambda_{i_j} \geq 0$ for $i_j < r$. For $i_j > r$ we must have $c^t v_{i_j} = 0$, by the construction of c in (2) and since r is the largest index to exit the simplicial subset. Applying these observations to (3.13), we get $c^t x \geq 0$ contradicting (3.14).

The rule of always exchanging smallest indices is the equivalent of Bland's rule in the simplex algorithm. It is remarkable that this simple rule is so powerful. In fact, the above algorithm furnishes an independent proof of Theorem 3.14 along with the substantial mathematical result: if $x \notin \text{cone}(V)$, there exists $c \in \mathbb{R}^d$ with $c^t x < 0$ and $c^t z \geq 0$ for $z \in \text{cone}(V)$. The latter will appear later under the name Farkas's lemma (Lemma 4.14) with a different proof.

3.7   Exercises

Exercise 3.1. Let $P = \{x \in \mathbb{R}^d \,|\, Ax \leq b\}$ be a polyhedron in $\mathbb{R}^d$. Prove that $(1 - t)x + ty \in P$ if $x, y \in P$ and $0 \leq t \leq 1$.

Exercise 3.2. Let $A, B \subseteq \mathbb{R}^d$ be convex subsets. Prove that $A \cap B$ is a convex subset. Give an example showing that $A \cup B$ does not have to be a convex subset.

Exercise 3.3. Prove in detail that an affine half space is a convex subset and that a polyhedron

$$P = \{x \in \mathbb{R}^d \,|\, Ax \leq b\}$$

is a convex subset of $\mathbb{R}^d$.

Exercise 3.4. Let A be a convex subset of $\mathbb{R}^d$. Prove that

$$A + z := \{x + z \,|\, x \in A\}$$

and

$$\lambda A := \{\lambda x \,|\, x \in A\}$$

are convex subsets of $\mathbb{R}^d$ for $z \in \mathbb{R}^d$ and $\lambda \in \mathbb{R}$. Let B be a convex subset of $\mathbb{R}^d$. Prove that

$$A + B := \{x + y \,|\, x \in A, \, y \in B\}$$

is a convex subset of $\mathbb{R}^d$.

Exercise 3.5. Let $v_1, v_2, v_3 \in \mathbb{R}^n$. Show that

$$\{(1 - \lambda)v_3 + \lambda((1 - \mu)v_1 + \mu v_2) \,|\, \lambda \in [0, 1], \mu \in [0, 1]\}$$
$$= \{\lambda_1 v_1 + \lambda_2 v_2 + \lambda_3 v_3 \,|\, \lambda_1, \lambda_2, \lambda_3 \geq 0, \ \lambda_1 + \lambda_2 + \lambda_3 = 1\}.$$

Exercise 3.6. Sketch the convex hull of

$$S = \{(0, 0), (1, 0), (1, 1)\} \subseteq \mathbb{R}^2.$$

Write $\text{conv}(S)$ as the intersection of three half planes.

Exercise 3.7. Let $u_1, u_2, v_1, v_2 \in \mathbb{R}^n$. Show that

$$\text{conv}(\{u_1, u_2\}) + \text{conv}(\{v_1, v_2\}) = \text{conv}(\{u_1 + v_1, u_1 + v_2, u_2 + v_1, u_2 + v_2\}).$$

Exercise 3.8. Let $S \subseteq \mathbb{R}^n$ be a convex subset and $v \in \mathbb{R}^n$. Show that

$$\{(1 - \lambda)s + \lambda v \mid \lambda \in [0, 1], s \in S\}$$

is a convex subset. Hint: compare with Exercise 3.5.

Exercise 3.9. Let $C = \text{conv}(\{v_1, \ldots, v_m\})$ and $P = C^\circ$ for $v_1, \ldots, v_m \in \mathbb{R}^d$. Prove that

$$P = \{\alpha \in \mathbb{R}^d \mid \alpha^t v_1 \leq 1, \ldots, \alpha^t v_m \leq 1\}$$

and that P is a polyhedron. Compute and sketch for C given in Example 3.6. Prove in general that P is bounded if 0 is an interior point of C.

Exercise 3.10. If $F \subseteq G \subseteq C$ are convex subsets of $\mathbb{R}^d$, prove that F is a face of C if F is a face of G and G is a face of C.

Exercise 3.11. Give an example of a convex subset $C \subseteq \mathbb{R}^d$ and a face $F \subseteq \mathbb{R}^d$, which is not exposed (hint: think about stretching a disc).

Exercise 3.12. Prove that $C \setminus F$ is a convex subset if F is a face of a convex subset C. Is it true that $F \subseteq C$ is a face if $C \setminus F$ is a convex subset?

Exercise 3.13. Let $X = \{x_1, \ldots, x_n\} \subseteq \mathbb{R}^d$.

(i) Prove that if $z \in \text{conv}(X)$ is an extreme point, then $z \in X$.
(ii) Suppose that $z \notin X$. Prove that z is an extreme point of $\text{conv}(\{z\} \cup X)$ if and only if

$$z \notin \text{conv}(X).$$

This means that the extreme points in a convex hull consists of the "non-redundant generators" (compare this with Figure 3.6).

Exercise 3.14. Prove in detail that

$$C = \{(x, y) \in \mathbb{R}^2 \mid x^2 + y^2 \leq 1\}$$

is a convex subset of $\mathbb{R}^2$. What are the extreme points of C? Can you prove it?

Exercise 3.15. Recall the notation

$$A + B = \{u + v \mid u \in A, v \in B\} \subseteq \mathbb{R}^n$$

for two subsets $A, B \subseteq \mathbb{R}^n$ and let $[u, v] := \text{conv}(\{u, v\})$ for $u, v \in \mathbb{R}^n$.

(i) Show that
$$[u, v] + \{w\} = [u + w, v + w]$$

for $u, v, w \in \mathbb{R}^n$.

(ii) Sketch
$$P = \left[\begin{pmatrix} 1 \\ 1 \end{pmatrix}, \begin{pmatrix} 1 \\ 2 \end{pmatrix}\right] + \left[\begin{pmatrix} 2 \\ 1 \end{pmatrix}, \begin{pmatrix} 3 \\ 2 \end{pmatrix}\right]$$

along with its extreme points in the plane.

(iii) Let
$$Q = P + \left[\begin{pmatrix} 3 \\ 1 \end{pmatrix}, \begin{pmatrix} 4 \\ 1 \end{pmatrix}\right].$$

Write Q as a convex hull of the minimal number of points and as an intersection of half planes.

(iv) Let A and B be convex sets and $u_0 \in A$ a point in A, which is *not* extreme. Show that $u_0 + b \in A + B$ is not extreme in $A + B$ for any $b \in B$.

(v) Show that
$$T = [x, y] + [z, w]$$

has at most 4 extreme points for $x, y, z, w \in \mathbb{R}^n$. Can T have 3 extreme points? 2?

(vi) Let $L_i = [u_i, v_i]$ for $i = 1, \ldots, m$, where $u_i, v_i \in \mathbb{R}^n$. Give an upper bound for how many extreme points

$$Z = L_1 + \cdots + L_m \qquad (3.15)$$

can have. Show that Z is the image of the unit cube $[0, 1]^m \subseteq \mathbb{R}^m$ under a suitable affine map.

The Minkowski sum of finitely many line segments (as in (3.15)) is called a *zonotope*.

Exercise 3.16. Give an example of a non-convex cone.

Exercise 3.17. Prove in detail that

$$C = \{(x, y, z) \in \mathbb{R}^3 \mid z \geq 0, \, x^2 + y^2 \leq z^2\}$$

is a convex cone. Is C finitely generated?

Exercise 3.18. Prove in detail that the recession cone $\mathrm{rec}(C)$ defined in (3.3) is a convex cone, where C is a convex subset. Perhaps the identity

$$x + (n + \lambda)d = (1 - \lambda)(x + nd) + \lambda(x + (n + 1)d)$$

might come in handy.

Exercise 3.19. What is the recession cone of a bounded convex subset?

Exercise 3.20. Can you give an example of an unbounded convex subset $C \subseteq \mathbb{R}^2$ with $\mathrm{rec}(C) = \{0\}$?

Exercise 3.21. Let

$$C = \mathrm{cone}\left(\left\{\begin{pmatrix}2\\1\end{pmatrix}, \begin{pmatrix}1\\2\end{pmatrix}\right\}\right).$$

(i) Show that

$$C^\circ = \mathrm{cone}\left(\left\{\begin{pmatrix}1\\-2\end{pmatrix}, \begin{pmatrix}-2\\1\end{pmatrix}\right\}\right).$$

(ii) Suppose that

$$C = \mathrm{cone}\left(\left\{\begin{pmatrix}a\\c\end{pmatrix}, \begin{pmatrix}b\\d\end{pmatrix}\right\}\right),$$

where

$$\begin{pmatrix}a & b\\c & d\end{pmatrix}$$

is an invertible matrix. How do you compute C°?

Exercise 3.22. The vector

$$v = \begin{pmatrix}7/4\\19/8\end{pmatrix}$$

is the convex combination

$$\tfrac{1}{8}\begin{pmatrix}1\\1\end{pmatrix} + \tfrac{1}{8}\begin{pmatrix}1\\2\end{pmatrix} + \tfrac{1}{4}\begin{pmatrix}2\\2\end{pmatrix} + \tfrac{1}{2}\begin{pmatrix}2\\3\end{pmatrix}$$

of four vectors in $\mathbb{R}^2$. Use the method outlined in Example 3.16 to answer the following questions.

(i) Is v in the convex hull of three of the four vectors?
(ii) Is v in the convex hull of two of the four vectors?

Exercise 3.23. Let C_1, C_2, C_3, C_4 be convex subsets of $\mathbb{R}^2$, such that any three of them have non-empty intersection i.e.,

$$C_1 \cap C_2 \cap C_3 \neq \emptyset$$
$$C_1 \cap C_2 \cap C_4 \neq \emptyset$$
$$C_1 \cap C_3 \cap C_4 \neq \emptyset$$
$$C_2 \cap C_3 \cap C_4 \neq \emptyset.$$

(i) Show that for $v_1, v_2, v_3, v_4 \in \mathbb{R}^2$ there are $\lambda_1, \lambda_2, \lambda_3, \lambda_4 \in \mathbb{R}$, not all zero, such that

$$\lambda_1 v_1 + \lambda_2 v_2 + \lambda_3 v_3 + \lambda_4 v_4 = 0$$
$$\lambda_1 + \lambda_2 + \lambda_3 + \lambda_4 = 0.$$

(ii) By assumption there exists

$$v_i \in \bigcap_{\substack{j=1 \\ j \neq i}}^{4} C_j$$

for $i = 1, \ldots, 4$. Suppose that with the notation in (i) we have

$$\lambda_1, \lambda_2 \geq 0$$
$$\lambda_3, \lambda_4 \leq 0.$$

Prove that

$$\frac{\lambda_1}{\lambda_1 + \lambda_2} v_1 + \frac{\lambda_2}{\lambda_2 + \lambda_2} v_2 \in C_1 \cap C_2 \cap C_3 \cap C_4.$$

(iii) Prove now in general that $C_1 \cap C_2 \cap C_3 \cap C_4 \neq \emptyset$.

(iv) Prove that if $C_1, \ldots, C_m$ are convex subsets of $\mathbb{R}^2$ where any three of them have non-empty intersection, then $C_1 \cap \cdots \cap C_m \neq \emptyset$.

(v) What is a natural generalization from $\mathbb{R}^2$ to $\mathbb{R}^n$ of the result in (iv)?

Exercise 3.24. Let S be a subset of $\mathbb{R}^n$ containing at least $n + 2$ points. Prove that there exists subsets $S_1, S_2 \subseteq S$, such that

(i) $S_1 \cap S_2 = \emptyset$

(ii) $S_1 \cup S_2 = S$

(iii) $\mathrm{conv}(S_1) \cap \mathrm{conv}(S_2) \neq \emptyset.$

Hint: write down an affine dependence between $v_1, \ldots, v_{n+2} \in S$. This result is called Radon's theorem.

Exercise 3.25. Use the result in Exercise 3.24 to give a complete proof of the natural generalization alluded to in Exercise 3.23 (v) (called Helly's theorem).

Exercise 3.26. Let e_1, e_2, e_3 denote the canonical basis vectors of $\mathbb{R}^3$ and let

$$C = \mathrm{conv}(\{e_1, -e_1, e_2, -e_2, e_3, -e_3\}).$$

Verify that

$$\left(\tfrac{1}{3}, \tfrac{1}{5}, \tfrac{1}{7}\right) \in C$$

by writing down a convex linear combination.

Chapter 4

Polyhedra

Unlike a planar disc (Figure 3.9) and the Lorentz cone (Figure 3.11), polyhedra are intersections of finitely many affine half spaces (see Figure 4.1).

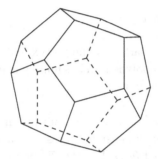

Figure 4.1: A (famous) polyhedron in $\mathbb{R}^3$.

The purpose of this chapter is to dig deeper into the structure of polyhedra only using Fourier-Motzkin elimination and basic linear algebra. A polyhedron has finitely many faces and they can be characterized using linear algebra. As a prelude to this characterization in Proposition 4.3, we will briefly mention the remarkable numerical relation known as Euler's formula. A proof of this formula is beyond the scope of this book. Fortunately it is one those mathematical miracles that can be appreciated almost instantly (see (4.1) and Figure 4.2).

A central result due to Minkowski[1] and Weyl[2] says that a polyhedron can be expressed as the Minkowski sum of a finitely generated cone and the convex hull of finitely many points. This result is conveniently treated in the setting of convex cones.

---

[1]Hermann Minkowski (1864–1909). German mathematician.
[2]Hermann Weyl (1886–1955). German mathematician.

A finitely generated convex cone is a polyhedron. It is not easy to write up a precise mathematical proof of this statement. Hermann Weyl, one of the great mathematicians of the 20th century, found it worthwhile to devote a paper [Weyl (1935)] to it. This result is now known as Weyl's theorem and was the subject of his last seminar in Göttingen in the summer of 1933 before he left for the United States.

We will give a deceptively short proof of Weyl's theorem using Fourier-Motzkin elimination. In all fairness we also enter into the ideas of Weyl's original proof in Chapter 5, hoping to convince the reader that it contains much deeper information on the minimal generators of a polyhedral cone.

The theory of polyhedra is an abundant source of interesting mathematics. We give a few examples in this chapter like Markov chains and steady states, duality in linear programming and doubly stochastic matrices.

4.1   Faces of polyhedra

Assume in the following remark that the convex hull of finitely many points is a polyhedron and that a polyhedron has finitely many faces. Both of these statements will be proved later.

Remark 4.1. If $C \subseteq \mathbb{R}^3$ is a convex subset, then its extreme points are called *vertices*, its one-dimensional faces are called *edges* and its two-dimensional faces are called *faces*. Let $P \subseteq \mathbb{R}^3$ be the convex hull of finitely many points and suppose that P is not contained in an affine plane. There is a remarkable numerical relation between the number of vertices V, the number of edges E and the number of faces F of P. This relation is

$$V - E + F = 2 \tag{4.1}$$

and seems to have been observed first by Descartes[3] in 1640. It was later used by Euler[4] in 1752 (see the chapter on topology in [Courant and Robbins (1941)]) and is today called Euler's formula for polyhedra. You should check (4.1) with the platonic solids depicted as wireframes in Figure 4.2 (for the cube we have $8 - 12 + 6 = 2$).

The formula (4.1) belongs to a rich branch of mathematics called polyhedral combinatorics. For the convex hull P of finitely many points in $\mathbb{R}^d$

---

[3]René Descartes (1596–1650). French mathematician.
[4]Leonhard Euler (1707–1783). Swiss mathematician.

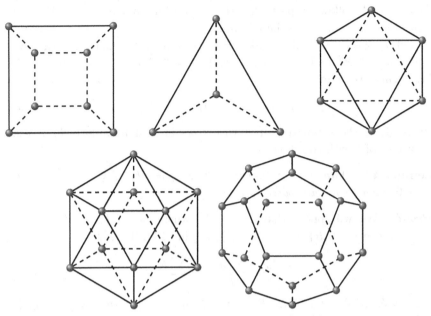

Figure 4.2: An illustration of Euler's formula (4.1) — the five so-called *platonic solids* with vertices and (in)visible edges marked: the *cube* (6 faces), the *tetrahedron* (4 faces), the *octahedron* (8 faces), the *icosahedron* (20 faces) and the *dodecahedron* (12 faces).

not contained in an affine hyperplane, (4.1) generalizes to the identity

$$f_0 - f_1 + f_2 - \cdots + (-1)^{d-1}f_{d-1} = 1 + (-1)^{d-1}, \qquad (4.2)$$

where f_j denotes the number of j-dimensional faces of P for $j = 0, 1, \ldots, d - 1$. For $d = 2$, (4.2) simply says that a convex polygon not contained in a line in $\mathbb{R}^2$ has as many vertices as edges. The generalization (4.2) is due to Poincaré[5] and was discovered by him in connection with his investigations of analysis situs (topology) around 1895.

For a proof of (4.2) see the more advanced texts [Barvinok (2002)], [Brøndsted (1983)] or [Ziegler (1995)]. There is a nice elementary treatment of (4.1) in Chapter V of [Courant and Robbins (1941)].

The faces of a polyhedron P are all exposed given as intersections with affine subspaces touching P. Algebraically, these affine subspaces are sets of solutions to linear equations arising from a subset of the linear inequalities

---

[5]Henri Poincaré (1854–1912). French mathematician.

defining P. It follows in particular that P has finitely many faces. We will prove these statements in the following.

Let $P = \{x \in \mathbb{R}^d \mid Ax \leq b\}$, where A is an $m \times d$ matrix with rows $a_1, \ldots, a_m \in \mathbb{R}^d$ and $b = (b_1, \ldots, b_m)^t \in \mathbb{R}^m$. For a given subset $I \subseteq \{1, \ldots, m\}$, we define

$$P_I := \{x \in P \mid A_I x = b_I\}, \tag{4.3}$$

where A_I is the submatrix of A with rows in I and similarly b_I is the subvector of b with coordinates in I.

Lemma 4.2. *With the notation in* (4.3) *suppose that* $I \subseteq \{1, \ldots, m\}$. *If* $P_I \neq \emptyset$, *then* P_I *is an exposed face of* P.

Proof.   We will prove that $P_I = F$, where $F = \{z \in P \mid \alpha^t z \leq \alpha^t x,$ for every $x \in P\}$ with $\alpha = -\sum_{i \in I} a_i$ in (3.1). If $x \in P$,

$$-\alpha^t x = \sum_{i \in I} a_i\, x \leq \sum_{i \in I} b_i.$$

If $P_I \neq \emptyset$, we therefore have $F = \{z \in P \mid \alpha^t z = \beta\}$ with $\beta = \sum_{i \in I} b_i$. It follows that $x \in P_I$ if and only if $x \in F$. $\square$

Let

$$I(z) = \{i \in \{1, \ldots, m\} \mid a_i z = b_i\}$$

for $z \in \mathbb{R}^d$ and put $A_z = A_{I(z)}$. In general, a face of a convex subset does not have to be exposed. Faces of polyhedra are exposed and we will prove that they all have the form (4.3).

Proposition 4.3. *If* F *is a non-empty face of* P, *then* $F = P_I$ *for some* $I \subseteq \{1, \ldots, m\}$. *If* P_I *is a non-empty face with* $I(z) = I$ *for some* $z \in P_I$, *then*

$$\dim P_I = d - \operatorname{rk} A_I.$$

Proof.   Let F be a non-empty face of P and $z \in F$ be chosen such that $I = I(z)$ has the minimal number of elements among $I(x)$, where $x \in F$. We claim that $F = P_I$. If $x \in P_I$, we let

$$z_\epsilon := z - \epsilon(x - z)$$

for $\epsilon > 0$. Since $a_j z = b_j$ if $j \in I$ and $a_j z < b_j$ if $j \notin I$, we have $z_\epsilon \in P_I$ for $\epsilon > 0$ sufficiently small. As

$$z = \frac{1}{1+\epsilon} z_\epsilon + \frac{\epsilon}{1+\epsilon} x$$

it follows by definition of a face that $x \in F$ and therefore $P_I \subseteq F$. Let us prove that $F \subseteq P_I$. For $y \in F$, we consider

$$z(\epsilon) := (1 - \epsilon)z + \epsilon y \in F$$

for $0 \le \epsilon \le 1$. To show that $y \in P_I$, we must prove that $I \subseteq I(y)$. Suppose on the contrary that there exists $j \in I \setminus I(y)$. Then $a_j y < b_j$ and we must have $I(z(\epsilon)) \subsetneq I(z)$ for ϵ sufficiently small. Therefore $|I(z(\epsilon))| < |I(z)|$ and this is in contradiction with the choice of z. Hence $F \subseteq P_I$.

The face P_I is contained in the affine subspace $\{x \in \mathbb{R}^d \mid A_I x = b_I\}$, which has dimension $r := d - \mathrm{rk}\, A_I$. To prove the stated dimension formula we need to find $r + 1$ affinely independent vectors in P_I. We may find r linearly independent vectors $v_1, \ldots, v_r \in \mathbb{R}^d$ with $A_I v_j = 0$ for $j = 1, \ldots, r$. For the given z we have $\mathrm{rk}\, A_z = \mathrm{rk}\, A_I$ and $a_j z < b_j$ if $j \notin I$. Therefore $z, z + \epsilon v_1, \ldots, z + \epsilon v_r$ are affinely independent vectors in P_I for $\epsilon > 0$ sufficiently small. $\square$

Definition 4.4. Let P be a polyhedron of dimension d and $F \subseteq P$ a face of P. If $\dim F = 0$, F is an extreme point also called a *vertex* of P. If $\dim F = 1$, F is called an *edge* of P. If $\dim F = d - 1$, F is called a *facet* of P.

An immediate consequence of Proposition 4.3 is the following corollary.

Corollary 4.5. *A polyhedron $P \subseteq \mathbb{R}^d$ has finitely many faces. A point $z \in P$ is a vertex if and only if A_z has rank d.*

Corollary 4.5 leads to a method for computing the vertices of a polyhedron $P \subseteq \mathbb{R}^d$. This is usually done traversing the subsets $I \subseteq \{1, \ldots, m\}$ with $|I| = d$ and A_I invertible. Here $x_0 = A_I^{-1} b_I$ is a vertex of P if and only if $x_0 \in P$. We will give a simple example illustrating the subscripting notation for faces and the computation of vertices.

Example 4.6. Let $Q = \{v \in \mathbb{R}^2 \mid Av \le b\}$ be given by

$$A = \begin{pmatrix} 1 & 0 \\ -1 & 0 \\ 0 & 1 \\ 0 & -1 \end{pmatrix} \quad \text{and} \quad b = \begin{pmatrix} 1 \\ 2 \\ 3 \\ 4 \end{pmatrix}.$$

Then

$$A_{\{1,2\}} = \begin{pmatrix} 1 & 0 \\ -1 & 0 \end{pmatrix} \quad \text{and} \quad b_{\{1,2\}} = \begin{pmatrix} 1 \\ 2 \end{pmatrix}.$$

Therefore $Q_{\{1,2\}} = \emptyset$. The subset $\{1\} \subseteq \{1,2,3,4\}$ defines an edge $Q_{\{1\}} = \{(x,y) \in \mathbb{R}^2 \,|\, x = 1, -4 \le y \le 3\}$ of Q.

Consider now $P = \{v \in \mathbb{R}^2 \,|\, Av \le b\}$, where

$$A = \begin{pmatrix} -1 & -1 \\ 2 & -1 \\ -1 & 2 \\ 1 & 2 \end{pmatrix} \quad \text{and} \quad b = \begin{pmatrix} 0 \\ 1 \\ 1 \\ 2 \end{pmatrix}.$$

For all of the 6 subsets $I \subseteq \{1,2,3,4\}$ with $|I| = 2$ one can check that A_I is invertible.

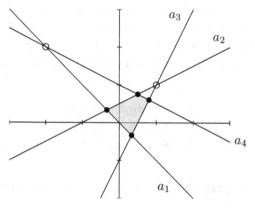

Figure 4.3: The polygon P has 4 vertices corresponding to the subsets $\{1,2\}, \{1,3\}, \{2,4\}$ and $\{3,4\}$ of the rows. The subsets $\{1,4\}$ and $\{2,3\}$ do not give rise to vertices of P.

However, P has only 4 vertices (see Figure 4.3). For $I = \{1,2\}$,

$$A_I = \begin{pmatrix} -1 & -1 \\ 2 & -1 \end{pmatrix} \quad \text{and} \quad b_I = \begin{pmatrix} 0 \\ 1 \end{pmatrix}.$$

Computing $A_I^{-1} b_I$ is the same as solving the system

$$-x - y = 0$$
$$2x - y = 1$$

of linear equations. This gives $z = (x,y) = (1/3, -1/3)$. Since $-1/3 + 2 \cdot (1/3) = 1/3 < 1$, $z \in P$ and z is a vertex of P. For the subset $J = \{2,3\}$,

$$A_J = \begin{pmatrix} 2 & -1 \\ -1 & 2 \end{pmatrix} \quad \text{and} \quad b_J = \begin{pmatrix} 1 \\ 1 \end{pmatrix}$$

and $A_J^{-1}b_J = (1, 1)^t$. Since $1 \cdot 1 + 2 \cdot 1 = 3 > 2$, this point violates the last inequality (the fourth row of A) in P. Therefore J does not give rise to a vertex of P.

4.2 Extreme points and linear optimization

A polyhedron does not necessarily contain a vertex. This is explained by the following result.

Theorem 4.7. *Let* $P = \{x \in \mathbb{R}^d \mid Ax \leq b\} \neq \emptyset$. *Then* $\mathrm{ext}(P) \neq \emptyset$ *if and only if* P *does not contain a line.*

Proof. If P contains a line $\{v + tu \mid t \in \mathbb{R}\}$ with $u \neq 0$, we must have $Au = 0$ (see Exercise 4.6). If $x \in F$, where F is a face if P, we have $\{x + tu \mid t \in \mathbb{R}\} \subseteq F$. This shows that $\dim F \geq 1$. Therefore $\mathrm{ext}(P) = \emptyset$.

Let $a_1, \ldots, a_m$ be the rows of A and $b = (b_1, \ldots, b_m)^t$. Assume that P does not contain a line and consider $z \in P$. If $\mathrm{rk}\, A_z = d$, then $z \in \mathrm{ext}(P)$. If not, $\mathrm{rk}\, A_z < d$ and we can find $u \neq 0$ with $A_z u = 0$. Since P does not contain a line we may assume that $a_j u > 0$ for some $j \notin I(z)$. Let

$$\lambda_0 = \max\{\lambda \mid a_j(z + \lambda u) \leq b_j, \text{ for every } j \notin I(z)\}$$
$$= \min_j \left\{ \frac{b_j - a_j z}{a_j u} \,\middle|\, a_j u > 0 \right\}$$

and

$$z_1 = z + \lambda_0 u.$$

Then $z_1 \in P$ and $\mathrm{rk}\, A_{z_1} > \mathrm{rk}\, A_z$. If $\mathrm{rk}\, A_{z_1} < d$ we continue this procedure with z_1. Eventually we will hit an extreme point. $\qquad\square$

The second part of the proof of Theorem 4.7 contains a constructive procedure for computing a vertex given a point $z \in P$. This procedure will be used later for computing a vertex (as part of the input for the simplex algorithm, see §5.4) in a polyhedron given by A and b. Let us illustrate it now with a simple example.

Example 4.8. Consider the polygon $P = \{x \in \mathbb{R}^2 \mid Ax \leq b\}$ with

$$A = \begin{pmatrix} -1 & -1 \\ 2 & -1 \\ -1 & 2 \end{pmatrix} \quad \text{and} \quad b = \begin{pmatrix} 0 \\ 1 \\ 1 \end{pmatrix}.$$

The point $z = (0,0) \in P$ is not a vertex, since $I(z) = \{1\}$, $A_z = (-1, -1)$ and rk $A_z = 1$. The vector $u = (1, -1)$ satisfies $A_z u = 0$. Tracing the steps in the proof of Theorem 4.7, we have $a_2^t u > 0, a_3^t u < 0$ and thus

$$\lambda_0 = \tfrac{1}{3}.$$

Therefore $z_1 = (\tfrac{1}{3}, -\tfrac{1}{3})$. Since $I(z_1) = \{1, 2\}$, rk $A_{z_1} = 2$ and $z_1 \in \text{ext}(P)$.

We have already seen a simple version of a linear optimization problem in Example 1.2. The general version has the form

$$\max\{c^t x \mid x \in P\}, \tag{4.4}$$

where $P \subseteq \mathbb{R}^d$ is a polyhedron and $c \in \mathbb{R}^d$. With our advanced knowledge of polyhedra we can now prove that the finitely many vertices of P play a central role in (4.4). The following result is a very important application of Theorem 4.7.

Corollary 4.9. *Let $P = \{x \in \mathbb{R}^d \mid Ax \leq b\}$ be a polyhedron with $\text{ext}(P) \neq \emptyset$ and $c \in \mathbb{R}^d$. If the optimization problem (4.4) has a maximum i.e., there exists $y \in P$ with $M = c^t y \geq c^t x$ for every $x \in P$, then this maximum is attained in a vertex:*

$$c^t x_0 = M$$

for some $x_0 \in \text{ext}(P)$.

Proof. The non-empty set

$$Q = \{x \in \mathbb{R}^d \mid c^t x = M\} \cap P$$

is a polyhedron not containing a line, since P does not contain a line by Theorem 4.7. Therefore Q contains an extreme point $z \in \text{ext}(Q)$ by Theorem 4.7. This is also an extreme point of P: if $z \notin \text{ext}(P)$, then rk $A_z < d$ and there exists a non-zero $u \in \mathbb{R}^d$ with $A_z u = 0$. For a small $\epsilon > 0$, $z \pm \epsilon u \in P$. Since $c^t z = M$, $c^t u = 0$ and therefore $z \pm \epsilon u \in Q$. The identity

$$z = \tfrac{1}{2}(z + \epsilon u) + \tfrac{1}{2}(z - \epsilon u)$$

shows that z is not an extreme point in Q. This is a contradiction. □

In §5.4 we will give an introduction to the simplex algorithm. This is a method for traversing the vertices along edges of P systematically improving the value of $c^t x$ in (4.4).

4.3 Weyl's theorem

In the beginning of this chapter we mentioned an important theorem due to Weyl. Weyl's theorem states that a finitely generated cone is polyhedral. To be in line with the notation $\{x \mid Ax \leq 0\}$ for a polyhedral cone, we will write a finitely generated convex cone

$$\text{cone}(\{v_1, \ldots, v_m\}) \subseteq \mathbb{R}^d$$

as

$$\{R\lambda \mid \lambda \geq 0\},$$

where R is the $n \times d$ matrix with columns $v_1, \ldots, v_m$. It is implicit here that $\lambda \in \mathbb{R}^m$.

Example 4.10. Consider a triangle T in $\mathbb{R}^2$ given as the convex hull

$$T = \text{conv}\left(\left\{\binom{1}{1}, \binom{3}{2}, \binom{2}{3}\right\}\right).$$

Then T is the intersection of 3 affine half spaces. Recall that $x \in T$ if and only if

$$\binom{x}{1} \in \left\{R\binom{\lambda_1}{\lambda_2 \atop \lambda_3} \,\middle|\, \lambda_1, \lambda_2, \lambda_3 \geq 0\right\} = \{R\lambda \mid \lambda \geq 0\}, \tag{4.5}$$

where

$$R = \begin{pmatrix} 1 & 3 & 2 \\ 1 & 2 & 3 \\ 1 & 1 & 1 \end{pmatrix}.$$

This is nothing but (3.4) phrased in matrix language. Here R is an invertible matrix and $x = R\lambda$ for $\lambda \geq 0$ if and only if $R^{-1}x = \lambda \geq 0$. With $A = -R^{-1}$, we have

$$Ax \leq 0 \iff x = R\lambda, \qquad \text{for } \lambda \geq 0.$$

Now you can use (4.5) and the matrix A to explicitly compute the affine half spaces defining T (how?).

The above example naturally leads to the following important definition.

Definition 4.11. Suppose that A is an $m \times d$ matrix and R a $d \times n$ matrix. Then (A, R) is called a *double description pair* if

$$\{x \in \mathbb{R}^d \mid Ax \leq 0\} = \{R\lambda \mid \lambda \in \mathbb{R}^n, \lambda \geq 0\}.$$

With this definition, Weyl's theorem can be stated.

Theorem 4.12 (Weyl). *For every matrix R, there exists a matrix A, such that (A, R) is a double description pair.*

A truly surprising application of Fourier-Motzkin elimination (Theorem 1.6) is in giving a short proof of Weyl's theorem rephrased in the lemma below.

Lemma 4.13. *Let $C = \{R\lambda \,|\, \lambda \geq 0\} \subseteq \mathbb{R}^d$ denote a finitely generated convex cone, where R is a $d \times n$ matrix for some $n \geq 1$. Then there exists an $m \times d$ matrix A for some $m \geq 1$, such that*

$$C = \{v \in \mathbb{R}^d \,|\, Av \leq 0\}.$$

Proof. Consider the polyhedral cone

$$P = \left\{ \begin{pmatrix} x \\ y \end{pmatrix} \;\middle|\; x \in \mathbb{R}^n, \, y \in \mathbb{R}^d, \, y = Rx, \, x \geq 0 \right\}$$

$$= \left\{ \begin{pmatrix} x \\ y \end{pmatrix} \;\middle|\; x \in \mathbb{R}^n, \, y \in \mathbb{R}^d, \, \begin{matrix} y - Rx \leq 0 \\ Rx - y \leq 0 \\ -x \leq 0 \end{matrix} \right\}$$

in $\mathbb{R}^{d+n}$. Notice that P is constructed so that $\pi(P) = C$, where $\pi : \mathbb{R}^{d+n} \to \mathbb{R}^d$ is the projection defined by $\pi(x, y) = y$. Since $\pi(\lambda x) = \lambda \pi(x)$ for $\lambda \geq 0$ and $x \in P$, C must be a cone and the existence of the matrix A follows by Theorem 1.6 and Proposition 3.9. $\square$

The proof above of Weyl's theorem only gives vague hints as how to compute A given R. You can go through the n steps of Fourier-Motzkin elimination from $\mathbb{R}^{d+n}$ to $\mathbb{R}^d$, but we will outline a more structured algorithm called the double description method in the next chapter.

The fact that there exists a matrix R, such that (A, R) is a double description pair given a matrix A is called Minkowski's theorem. It comes as a result of Weyl's theorem and a clever duality operation based on a very important result called Farkas's lemma.

4.4   Farkas's lemma

The lemma of Farkas[6] is an extremely important result in the theory of polyhedra. Farkas published his result in the late 19th century. We will show that Farkas's lemma is a consequence of Weyl's theorem and give several applications including a proof of Minkowski's theorem and Gordan's theorem.

---

[6]Gyula Farkas (1847–1930), Hungarian mathematician.

Lemma 4.14 (Farkas). *Let A be an $m \times n$ matrix and $b \in \mathbb{R}^m$. Then precisely one of the following two conditions hold.*

(1) *The system $Ax = b$ of linear equations is solvable with $x \geq 0$.*
(2) *There exists $y \in \mathbb{R}^m$ such that*

$$y^t A \leq 0 \quad and \quad y^t b > 0.$$

Proof. Both conditions cannot hold at the same time, since

$$y^t b = y^t(Ax) = (y^t A)x \leq 0$$

if $y^t A \leq 0$ and $x \geq 0$. This inequality contradicts $y^t b > 0$. The surprise is that the second condition holds if the first does not: let C denote the cone in $\mathbb{R}^m$ generated by the column vectors of A. By Lemma 4.13 there exists a matrix Y, such that

$$C = \{v \in \mathbb{R}^m \mid Yv \leq 0\}.$$

If the first condition does not hold, then $b \notin C$. This means that $Yb \not\leq 0$ i.e., there exists a row vector y in Y, such that $y^t b > 0$. If a is a column vector in A, we must have $y^t a \leq 0$ by definition of C. Therefore $y^t A \leq 0$ and the proof is complete. $\qquad\square$

Farkas's original proof of Lemma 4.14 was rather long and complicated. As is often the case with important mathematical results, its proof has been substantially simplified (compare also with the algorithmic proof given in §3.6).

Example 4.15. Does the system

$$
\begin{aligned}
x + 2y - z + 3w &= -1 \\
-x + y - 2z + w &= 1 \\
x + y + 3z - w &= -1
\end{aligned}
\tag{4.6}
$$

of linear equations, have a solution with $x, y, z, w \geq 0$? A certificate that this is impossible is the vector $y = (-1, 0, -1)^t \in \mathbb{R}^3$, since

$$y^t \begin{pmatrix} 1 & 2 & -1 & 3 \\ -1 & 1 & -2 & 1 \\ 1 & 1 & 3 & -1 \end{pmatrix} = (-2, -3, -2, -2) \leq 0$$

whereas

$$y^t \begin{pmatrix} -1 \\ 1 \\ -1 \end{pmatrix} = 2 > 0.$$

The system of equations (4.6) has many solutions (find one!) but not one without a negative coordinate.

4.5   Three applications of Farkas's lemma

4.5.1   *Markov chains and steady states*

In discrete probability one studies a trial with a finite set of outcomes (events) say $E_1, \ldots, E_n$ and associated probabilities $p(E_1), \ldots, p(E_n) \geq 0$ with $p(E_1) + \cdots + p(E_n) = 1$. Here the trial could be flipping a coin and the events head or tail. If the trials are independent, the probability of having the outcome $E_{i_1}, \ldots, E_{i_m}$ in m successive trials is

$$p(E_{i_1}, \ldots, E_{i_m}) = p(E_{i_1}) \cdots p(E_{i_m}). \qquad (4.7)$$

In tossing a fair coin one would for example expect that the probability of four heads in a row is $(\frac{1}{2})^4$. It is important to emphasize that the probability of the coin turning up heads in the fourth trial is still $\frac{1}{2}$ independently of observing three heads in the previous trials. It is assumed the universe has no memory of your past luck (or misfortune). This independence condition is not assumed for Markov chains.

In a Markov[7] chain one permits the outcome of a trial to depend on the previous trial (see Chapter XV of [Feller (1957)]). A *Markov chain* on $E_1, \ldots, E_n$ is given by n probabilities $a_1, \ldots, a_n$ of an initial trial along with n^2 transition probabilities p_{ij} for $i, j = 1, \ldots, n$. Here p_{ij} is interpreted as the conditional probability $p(E_j \mid E_i)$ that E_j occurs in the next trial given that E_i has occurred in the current trial.

Let $p^{(m)}(E_j)$ denote the probability of E_j in the m-th trial of a Markov chain with the convention that $p^{(0)}(E_j) = a_j$. Then we have the fundamental formula

$$p^{(m)}(E_j) = p_{1j} p^{(m-1)}(E_1) + \cdots + p_{nj} p^{(m-1)}(E_n) \qquad (4.8)$$

by properties of conditional probabilities (in the model of independent trials we would simply have $p^{(m)}(E_j) = p(E_j)$). In matrix language (4.8) reads

$$\begin{pmatrix} p^{(m)}(E_1) \\ \vdots \\ p^{(m)}(E_n) \end{pmatrix} = P \begin{pmatrix} p^{(m-1)}(E_1) \\ \vdots \\ p^{(m-1)}(E_n) \end{pmatrix} \qquad (4.9)$$

---

[7]Andrei Markov (1856–1922). Russian mathematician.

with

$$P = \begin{pmatrix} p_{11} & \cdots & p_{n1} \\ \vdots & \ddots & \vdots \\ p_{1n} & \cdots & p_{nn} \end{pmatrix}. \qquad (4.10)$$

In particular we have the formula

$$\begin{pmatrix} p^{(m)}(E_1) \\ \vdots \\ p^{(m)}(E_n) \end{pmatrix} = P^m \begin{pmatrix} a_1 \\ \vdots \\ a_n \end{pmatrix}. \qquad (4.11)$$

One of the miracles of Markov chains is that the behavior of $P^{(m)}(E_j)$ for m large does not depend on the initial probabilities $a_1, \ldots, a_n$ if the transition probabilities p_{ij} are all positive. In precise terms,

$$\lim_{m \to \infty} P^{(m)}(E_j)$$

exists and is independent of $a_1, \ldots, a_n$ if $P > 0$. In view of (4.11), this means that successive powers

$$P, P^2, P^3, \ldots$$

of the matrix in (4.10) must converge to a matrix with identical columns (of probability vectors). As a reality check of this statement, you can easily carry out experiments like

$$\begin{pmatrix} 0.8 & 0.6 \\ 0.2 & 0.4 \end{pmatrix}, \begin{pmatrix} 0.76 & 0.72 \\ 0.24 & 0.28 \end{pmatrix}, \begin{pmatrix} 0.752 & 0.744 \\ 0.248 & 0.256 \end{pmatrix}, \begin{pmatrix} 0.7504 & 0.7488 \\ 0.2496 & 0.2512 \end{pmatrix}, \cdots$$
$$(4.12)$$

In light of this limiting property and (4.9) we call a vector $x = (p_1, \ldots, p_n) \geq 0$ with $p_1 + \cdots + p_n = 1$ a *steady state* for a Markov chain if $Px = x$.

Example 4.16. As a concrete example of a Markov chain we consider a very simplified model of consumer behavior in the cell phone market with two states: S_1 signifying that the customer has a cell phone and S_2 signifying that the customer does not have a cell phone. If a customer does not have a cell phone, the probability that she will buy one within a year is 60 %. The probability that a customer with a cell phone will decide to get rid of it and sell it is 20 %. We will not need the initial probabilities in this case i.e., the percentage of people owning a cell phone in the first year. The specified data give

$$P = \begin{pmatrix} 0.8 & 0.6 \\ 0.2 & 0.4 \end{pmatrix}$$

as the matrix of transition probabilities (see Figure 4.4).

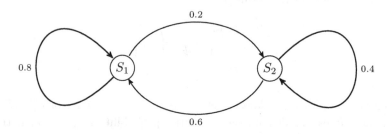

Figure 4.4: The matrix P of transition probabilities in the cell phone Markov chain showed as a labeled graph.

The steady state for this Markov chain can be read off from the computational experiment in (4.12). It seems that in the long run 75 % will have a cell phone and 25 % will not. This can also be proved rigorously (see Exercise 4.8).

The existence of a steady state for a Markov chain can be inferred from Farkas's lemma. In fact, this existence may be reformulated as the problem: does

$$\begin{pmatrix} p_{11} - 1 & p_{21} & \cdots & p_{n1} \\ p_{12} & p_{22} - 1 & \cdots & p_{n2} \\ \vdots & \vdots & \ddots & \vdots \\ p_{1n} & p_{2n} & \cdots & p_{nn} - 1 \\ 1 & 1 & \cdots & 1 \end{pmatrix} \begin{pmatrix} p_1 \\ \vdots \\ p_n \end{pmatrix} = \begin{pmatrix} 0 \\ \vdots \\ 0 \\ 1 \end{pmatrix} \tag{4.13}$$

have a solution $(p_1, \ldots, p_n) \geq 0$? We will prove this using Lemma 4.14 by showing that the second alternative cannot hold. In the case of (4.13), disproving the second alternative means showing that there cannot exist $y = (y_1, \ldots, y_n, y_{n+1})^t \in \mathbb{R}^{n+1}$, such that

$$(p_{11} - 1)y_1 + p_{12}y_2 + \cdots + p_{1n}y_n + y_{n+1} \leq 0$$
$$p_{21}y_1 + (p_{22} - 1)y_2 + \cdots + p_{2n}y_n + y_{n+1} \leq 0$$
$$\vdots$$
$$p_{n1}y_1 + p_{n2}y_2 + \cdots + (p_{nn} - 1)y_n + y_{n+1} \leq 0$$
$$y_{n+1} > 0.$$

If y exists satisfying the above inequalities we can conclude that

$$p_{11}y_1 + \cdots + p_{1n}y_n < y_1$$
$$p_{21}y_1 + \cdots + p_{2n}y_n < y_2$$
$$\vdots \qquad\qquad (4.14)$$
$$p_{n1}y_1 + \cdots + p_{nn}y_n < y_n.$$

Suppose that $y_k = \min(y_1, \ldots, y_n)$. Then

$$y_k \leq p_{j1}y_1 + p_{j2}y_2 + \cdots + p_{jn}y_n,$$

since $p_{j1} + \cdots + p_{jn} = 1$ for every $j = 1, \ldots, n$. This contradicts the k-th inequality in (4.14). Therefore the second alternative does not hold and the first alternative proves the existence of a steady state.

4.5.2 *Gordan's theorem*

Another application of Farkas's lemma is in the short proof of Gordan's theorem below. The proof is very much related to the trick used in proving the existence of a steady state for a Markov chain.

We will prove the following classical result [Gordan (1873)] due to Gordan[8] dating back to 1873:

Theorem 4.17. *Let A be an $m \times n$ matrix. Then precisely one of the following two conditions holds.*

(1) *There exists $x \in \mathbb{R}^n$, such that*

$$Ax > 0.$$

(2) *There exists a non-zero $y \geq 0$ in $\mathbb{R}^m$ such that*

$$y^t A = 0.$$

Proof. Suppose that $A = (a_{ij})$ for $i = 1, \ldots, m$ and $j = 1, \ldots, n$. Then $y^t A = 0$ has a non-zero solution $y \in \mathbb{R}^m$ with $y \geq 0$ if and only if

$$a_{11}y_1 + a_{21}y_2 + \cdots + a_{m1}y_m = 0$$
$$\vdots$$
$$a_{1n}y_1 + a_{2n}y_2 + \cdots + a_{mn}y_m = 0$$
$$y_1 + \cdots + y_m = 1$$

---

[8]Paul Gordan (1837–1912), German mathematician.

has a solution $y \geq 0$. The Farkas alternative (the second condition in Lemma 4.14) to this set of equations is

$$a_{11}x_1 + \cdots + a_{1n}x_n + z \leq 0$$
$$\vdots$$
$$a_{m1}x_1 + \cdots + a_{mn}x_n + z \leq 0$$
$$z > 0.$$

This is the same as $x = (x_1, \ldots, x_n)^t$ satisfying $Ax < 0$.   □

In Theorem 10.6, Gordan's theorem will play a crucial role in deriving the Karush-Kuhn-Tucker conditions for optimization problems.

4.5.3   *Duality in linear programming*

Perhaps the most striking application of Farkas's lemma is in linear optimization as encountered in Example 1.2 and §4.2. A linear program in *standard form* refers to a linear optimization problem of the form

$$\min\{c^t x \mid Ax = b, x \geq 0\}, \tag{4.15}$$

where A is an $m \times d$ matrix, $b \in \mathbb{R}^m$ and $c \in \mathbb{R}^d$. An *optimal solution* to (4.15) is a vector $x_0 \in \mathbb{R}^d$ with $x_0 \geq 0$, $Ax_0 = b$ and $c^t x_0 \leq c^t x$ for every $x \in \mathbb{R}^d$ with $x \geq 0$ and $Ax = b$.

There is a rather miraculous connection between (4.15) and the linear optimization problem

$$\max\{y^t b \mid y^t A \leq c^t\}. \tag{4.16}$$

An optimal solution to (4.16) is a vector $y_0 \in \mathbb{R}^m$ with $y_0^t A \leq c^t$ and $y^t b \leq y_0^t b$ for every $y \in \mathbb{R}^m$ with $y^t A \leq c^t$. We will refer to (4.15) as the *primal problem* and (4.16) as the *dual problem*. The surprising link between the primal and dual problems is called the duality theorem for linear programming (Theorem 4.19) and represents a subtle application of Farkas's lemma. The duality theorem was first envisioned by John von Neumann[9] in a working paper from 1947 at the Institute for Advanced Study. We first give a preliminary result also involving Farkas's lemma.

Lemma 4.18. *If* $\min\{c^t x \mid Ax = b, \ x \geq 0\}$ *has an optimal solution* x_0, *there exists* $y \in \mathbb{R}^m$ *such that*

$$y^t A \leq c^t. \tag{4.17}$$

---

[9]John von Neumann (1903–1957). Hungarian-American mathematician.

Similarly if $\max\{y^t b \mid y^t A \leq c^t\}$ *has an optimal solution* y_0, *there exists* $x \in \mathbb{R}^d$ *with* $x \geq 0$ *and* $Ax = b$.

Proof. We will convert the question of solutions to (4.17) into a form suitable for Farkas's lemma. Adjoining the variables $y^+, y^- \in \mathbb{R}^m$ and $z \in \mathbb{R}^d$ with $y^+, y^-, z \geq 0$, (4.17) becomes $(y^+ - y^-)^t A + z = c^t$ assuming that $y = y^+ - y^-$. In matrix language this can be written as

$$\left(A^t \mid -A^t \mid I \right) \begin{pmatrix} y^+ \\ y^- \\ z \end{pmatrix} = c, \tag{4.18}$$

where I is the $d \times d$ identity matrix. If (4.18) is unsolvable with $y^+, y^-, z \geq 0$, there exists $\alpha \in \mathbb{R}^d$ with $A\alpha = 0$, $\alpha \geq 0$ and $c^t \alpha < 0$ by Farkas's lemma. This implies

$$A(x_0 + \alpha) = b, \quad x_0 + \alpha \geq 0, \quad \text{and} \quad c^t(x_0 + \alpha) < c^t x_0$$

contradicting the optimality of x_0. Similarly if $Ax = b$ is unsolvable with $x \geq 0$ there exists $\beta \in \mathbb{R}^m$ with $\beta^t A \leq 0$ and $\beta^t b > 0$. This implies $(y_0 + \beta)^t A \leq c^t$ and $(\beta + y_0)^t b > y_0^t b$ contradicting the optimality of y_0. $\square$

Theorem 4.19 (Linear programming duality). *Let A be an $m \times d$ matrix, $b \in \mathbb{R}^m$ and $c \in \mathbb{R}^d$. If there exist $x_0 \geq 0$ with $Ax_0 = b$ and y_0 with $y_0^t A \leq c^t$, then*

$$\max\{y^t b \mid y^t A \leq c^t\} = \min\{c^t x \mid x \geq 0, Ax = b\}.$$

The optimization problem $\min\{c^t x \mid x \geq 0, Ax = b\}$ has an optimal solution if and only if the dual optimization problem $\max\{y^t b \mid y^t A \leq c^t\}$ has an optimal solution. In this case their optimal values are equal.

Proof. Notice first that if $x \geq 0$, $Ax = b$ and $y^t A \leq c^t$, then

$$y^t b = y^t(Ax) = (y^t A)x \leq c^t x. \tag{4.19}$$

It is therefore enough to prove that

$$\begin{aligned} Ax &= b \\ y^t A &\leq c^t \\ y^t b &= c^t x \end{aligned} \tag{4.20}$$

has a solution with $x \in \mathbb{R}^d, y \in \mathbb{R}^m$ and $x \geq 0$. To convert this system of inequalities for use in Farkas's lemma, we introduce $y^+, y^- \in \mathbb{R}^m$, $z \in \mathbb{R}^d$ so that (4.20) having a solution is equivalent to

$$
\begin{pmatrix} A & 0 & 0 & 0 \\ 0 & A^t & -A^t & I \\ c^t & -b^t & b^t & 0 \end{pmatrix} \begin{pmatrix} x \\ y^+ \\ y^- \\ z \end{pmatrix} = \begin{pmatrix} b \\ c \\ 0 \end{pmatrix} \tag{4.21}
$$

having a solution with $(x, y^+, y^-, z) \geq 0$, where $x, z \in \mathbb{R}^d$ and $y^+, y^- \in \mathbb{R}^m$. Now we are in a position to apply Farkas. If (4.21) does not have a solution, Farkas implies the existence of $\alpha \in \mathbb{R}^m, \beta \in \mathbb{R}^d$ and $\gamma \in \mathbb{R}$, such that

$$
\alpha^t A + \gamma c^t \geq 0 \tag{4.22}
$$

$$
\beta^t A^t = \gamma b^t \tag{4.23}
$$

$$
\beta \geq 0 \tag{4.24}
$$

$$
\alpha^t b + \beta^t c < 0. \tag{4.25}
$$

If $\gamma > 0$, then $x = \beta/\gamma$ satisfies $Ax = b$ and $x \geq 0$ by (4.23) and (4.24). Similarly $y = -\alpha/\gamma$ satisfies $y^t A \leq c^t$ by (4.22). Dividing (4.25) by γ implies that $c^t x < y^t b$ contradicting (4.19). Therefore $\gamma \leq 0$.

By assumption there exists $x_0 \geq 0$ with $Ax_0 = b$ and y_0 with $y_0^t A \leq c^t$. Multiplying (4.22) from the right by x_0 gives $\alpha^t b \geq -\gamma c^t x_0$. Multiplying (4.23) from the right by y_0 gives $\beta^t c \geq \gamma b^t y_0 = \gamma y_0^t b$. By (4.25) we get

$$
0 > \alpha^t b + \beta^t c \geq \gamma(y_0^t b - c^t x_0)
$$

implying $\gamma > 0$, since $y_0^t b - c^t x_0 \leq 0$ by (4.19). This contradicts $\gamma \leq 0$. The last claim in the theorem now follows from Lemma 4.18. □

Let us see Theorem 4.19 applied in a concrete example.

Example 4.20. Consider the linear program

$$
\min \quad x_1 + 2x_2 + 3x_3
$$

subject to

$$
\begin{aligned}
x_1 - 2x_2 + x_3 &= 4 \\
-x_1 + 3x_2 \quad\;\; &= 5
\end{aligned}
$$

and $x_1 \geq 0$, $x_2 \geq 0$, $x_3 \geq 0$. This is in standard form and the dual problem is

$$\text{max} \quad 4y_1 + 5y_2$$
$$\text{subject to}$$
$$
\begin{aligned}
y_1 \;-\;\; y_2 &\leq 1 \\
-2y_1 \;+\; 3y_2 &\leq 2 \\
y_1 \qquad\quad &\leq 3 .
\end{aligned}
$$

Here you can explicitly sketch the dual problem in the plane (or use Fourier-Motzkin elimination as in Example 1.2) and conclude that the optimal solution is $y_1 = 3$ and $y_2 = 8/3$ with optimal value $76/3$. In general if y_0 is an optimal solution to the dual and x_0 is an optimal solution to the primal problem, then

$$c^t x_0 = y_0^t b = y_0^t A x_0.$$

This implies $(c^t - y_0^t A) x_0 = 0$ and puts some restrictions on x_0, since both $c^t - y_0^t A \geq 0$ and $x_0 \geq 0$. If $c_j^t - y_0^t A_j > 0$, where A_j is the j-th column in A, then the j-th coordinate in x_0 must be 0. This phenomenon is called *complementary slackness* – the same coordinates of $c^t - y_0^t A$ and x_0 cannot both be non-zero simultaneously. In our example

$$c^t - y_0^t A = (1,\, 2,\, 3) - (3,\, \tfrac{8}{3}) \begin{pmatrix} 1 & -2 & 1 \\ -1 & 3 & 0 \end{pmatrix} = (\tfrac{2}{3},\, 0,\, 0)$$

showing that $x_1 = 0$ in the optimal solution of the primal problem. With this information, it follows that

$$
\begin{aligned}
-2x_2 \;+\; x_3 &= 4 \\
3x_2 \qquad\;\; &= 5
\end{aligned}
$$

and $x_1 = 0, x_2 = 5/3$ and $x_3 = 22/3$. You can verify that $x_1 + 2x_2 + 3x_3 = 76/3$. Therefore we must have an optimal solution.

Usually linear programs in standard form are solved algorithmically using the simplex algorithm, which is a systematic way of improving $c^t x$ touring the vertices of the polyhedron $P = \{x \in \mathbb{R}^d \mid Ax = b, x \geq 0\}$. The simplex algorithm is arguably among the most applied algorithms in the history of mathematics. It is extremely surprising that this method has proven so effective (the polyhedron P may have an astronomical number of extreme points!). We will explain the precise steps of the simplex method for polyhedra in general form in §5.4.

We now show how Farkas's lemma gives rise to a very clever duality operation for double description pairs. This operation is used in deducing Minkowski's theorem from Weyl's theorem (Theorem 4.12).

4.6   Minkowski's theorem

Lemma 4.21. *If (A, R) is a double description pair, then (R^t, A^t) is a double description pair.*

Proof.   We start out with the assumption that

$$\{x \mid Ax \le 0\} = \{R\lambda \mid \lambda \ge 0\} \tag{4.26}$$

and wish to prove that

$$\{x \mid R^t x \le 0\} = \{A^t \lambda \mid \lambda \ge 0\}. \tag{4.27}$$

To prove the inclusion $\supseteq$ in (4.27), notice by (4.26) that $Ar \le 0$ if r is a column vector of R. Therefore AR is matrix with entries ≤ 0 and $x = A^t \lambda$ for $\lambda \ge 0$ implies that

$$R^t x = R^t A^t \lambda = (AR)^t \lambda \le 0.$$

To prove the inclusion $\subseteq$ in (4.27) suppose that $R^t x \le 0$ and there does not exist $\lambda \ge 0$, such that $x = A^t \lambda$. Then Lemma 4.14 assures the existence of a vector y with $y^t x > 0$ and $y^t A^t \le 0$. This implies $Ay \le 0$ and therefore $y = R\lambda$ for some $\lambda \ge 0$ by (4.26). But $(R\lambda)^t x = \lambda^t R^t x \le 0$ contradicting that $y^t x > 0$. Therefore $x = A^t \lambda$ for some $\lambda \ge 0$ finishing the proof that (R^t, A^t) is a double description pair. $\square$

We are now ready to prove the "converse" of Weyl's theorem stating that polyhedral cones are finitely generated.

Theorem 4.22 (Minkowski). *For every matrix A, there exists a matrix R, such that (A, R) is a double description pair.*

Proof.   By Theorem 4.12, there exists a matrix S so that (S, A^t) is a double description pair. Putting $R = S^t$, it follows by Lemma 4.21 that (A, R) is a double description pair. $\square$

4.7 Parametrization of polyhedra

Fourier-Motzkin elimination may be viewed as a parametrization method for polyhedra. From the description of a polyhedron P as the set of solutions to a system of linear inequalities it is not obvious how to find an explicit point $x \in P$ or even decide if $P = \emptyset$. Projecting P onto the different axes, Fourier-Motzkin elimination provides a method of building up P. The general result is given in this section. It basically says that a polyhedron is the Minkowski sum of a finitely generated cone and a polytope.

This result is a consequence of Theorem 4.12 (Weyl's theorem) and Theorem 4.22 (Minkowski's theorem) telling us that finitely generated cones and polyhedral cones are two ways of describing the same objects.

To establish this link we relate a polyhedron $P \subseteq \mathbb{R}^d$ to a convex cone $\hat{P} \subseteq \mathbb{R}^{d+1}$, such that

$$x \in P \iff \begin{pmatrix} x \\ 1 \end{pmatrix} \in \hat{P}.$$

In the words of Herman Weyl: "Man geht zweckmäßig von der *homogenen* Formulierung aus."

Theorem 4.23 (Minkowski and Weyl). *A subset $P \subseteq \mathbb{R}^n$ is a polyhedron if and only if*

$$P = C + Q,$$

where C is a finitely generated cone and Q a polytope.

Proof. Suppose first that $P = \{x \in \mathbb{R}^n \mid Ax \leq b\}$. Then

$$\hat{P} = \left\{ \begin{pmatrix} x \\ z \end{pmatrix} \,\middle|\, x \in \mathbb{R}^n,\, z \in \mathbb{R},\, Ax \leq bz,\, z \geq 0 \right\}$$

is a polyhedral cone in $\mathbb{R}^{n+1}$ with the property that

$$x \in P \iff \begin{pmatrix} x \\ 1 \end{pmatrix} \in \hat{P}. \tag{4.28}$$

By Theorem 4.22, $\hat{P}$ is a finitely generated cone. Suppose that

$$\hat{P} = \text{cone}\left(\left\{ \begin{pmatrix} x_1 \\ z_1 \end{pmatrix}, \ldots, \begin{pmatrix} x_m \\ z_m \end{pmatrix} \right\} \right),$$

where we may assume that z_j is 0 or 1. Let C be the cone generated by x_i with $z_i = 0$ and Q the convex hull of x_i with $z_i = 1$. By (4.28) we can conclude that $P = C + Q$.

Suppose on the other hand that $P = C + Q$, where $C = \text{cone}(\{u_1, \ldots, u_r\})$ and $Q = \text{conv}(\{v_1, \ldots, v_s\})$. By Theorem 4.12,

$$\hat{P} = \text{cone}\left(\left\{\binom{u_1}{0}, \ldots, \binom{u_r}{0}, \binom{v_1}{1}, \ldots, \binom{v_s}{1}\right\}\right)$$
$$= \left\{\binom{x}{z} \,\middle|\, x \in \mathbb{R}^n,\, z \in \mathbb{R},\, [A \mid b]\binom{x}{z} \leq 0\right\}$$
$$= \left\{\binom{x}{z} \,\middle|\, x \in \mathbb{R}^n,\, z \in \mathbb{R},\, Ax + bz \leq 0\right\},$$

where A is an $m \times n$ matrix and $b \in \mathbb{R}^m$. This implies that

$$P = \{x \in \mathbb{R}^n \mid Ax \leq -b\},$$

again by applying (4.28). □

It is not too hard to verify that the characteristic cone of a polyhedron $P = \{x \in \mathbb{R}^n \mid Ax \leq b\}$ is the polyhedral cone

$$\text{rec}(P) = \{x \in \mathbb{R}^n \mid Ax \leq 0\}.$$

We have the following refinement of Theorem 4.23 for polyhedra containing extreme points.

Theorem 4.24. *Let* $P = \{x \in \mathbb{R}^n \mid Ax \leq b\}$ *be a polyhedron with* $\text{ext}(P) \neq \emptyset$. *Then*
$$P = \text{conv}(\text{ext}(P)) + \text{rec}(P).$$

Proof. Let $Q = \text{conv}(\text{ext}(P)) + \text{rec}(P)$. The definition of $\text{rec}(P)$ shows that $Q \subseteq P$. Suppose that $z \in P \setminus Q$. Since Q is a polyhedron by Theorem 4.23, there exists $c \in \mathbb{R}^n$ with $c^t z > c^t x$ for every $x \in Q$. This shows that $c^t d \leq 0$ for every $d \in \text{rec}(P)$. Writing P as the sum of a polytope $\text{conv}(\{v_1, \ldots, v_r\})$ and a convex cone C again by Theorem 4.23, it follows that $C \subseteq \text{rec}(P)$. With $M = \max\{c^t v_1, \ldots, c^t v_r\}$ we therefore have $M \geq c^t x$ for every $x \in P$ and there exists $x_0 \in \{v_1, \ldots, v_r\} \subseteq P$, such that $M = c^t x_0 \geq c^t x$ for every $x \in P$. In particular $c^t x_0 \geq c^t z > c^t x$ for every $x \in \text{ext}(P) \subseteq Q$ in contradiction with Corollary 4.9. □

4.8   Doubly stochastic matrices: The Birkhoff polytope

A *doubly stochastic matrix* is a square matrix with non-negative entries such that the entries in every row and every column sum to one. The set

B_n of doubly stochastic $n \times n$ matrices is a polyhedron in $\mathbb{R}^{n^2}$ given by

$$x_{ij} \geq 0, \qquad \text{for } i, j = 1, \ldots, n \qquad (4.29)$$

$$x_{i1} + \cdots + x_{in} = 1, \qquad \text{for } i = 1, \ldots, n \qquad (4.30)$$

$$x_{1j} + \cdots + x_{nj} = 1, \qquad \text{for } j = 1, \ldots, n. \qquad (4.31)$$

The entries in a doubly stochastic matrix are bounded and B_n is a polytope by Theorem 4.23. This polytope is called the *Birkhoff polytope*. One can prove that $\dim B_n = (n-1)^2$ (see Exercise 4.11).

A *permutation matrix* is a doubly stochastic matrix with entries in $\{0, 1\}$ i.e., with precisely one entry $= 1$ in every row and column. A remarkable fact is that any doubly stochastic matrix is a convex linear combination of these matrices. This was proved by Birkhoff[10] in 1946 and von Neumann in 1953. Birkhoff considered the problem in the context of linear algebra and von Neumann in the context of matrix games.

Example 4.25. In B_3,

$$\begin{pmatrix} 0.4 & 0.2 & 0.4 \\ 0.2 & 0.5 & 0.3 \\ 0.4 & 0.3 & 0.3 \end{pmatrix} = 0.3 \begin{pmatrix} 1 & 0 & 0 \\ 0 & 0 & 1 \\ 0 & 1 & 0 \end{pmatrix} + 0.1 \begin{pmatrix} 1 & 0 & 0 \\ 0 & 1 & 0 \\ 0 & 0 & 1 \end{pmatrix}$$

$$+ 0.4 \begin{pmatrix} 0 & 0 & 1 \\ 0 & 1 & 0 \\ 1 & 0 & 0 \end{pmatrix} + 0.2 \begin{pmatrix} 0 & 1 & 0 \\ 1 & 0 & 0 \\ 0 & 0 & 1 \end{pmatrix}.$$

Theorem 4.26 (Birkhoff, von Neumann). *A doubly stochastic matrix is a convex linear combination of permutation matrices.*

Proof. We know by Theorem 4.23 that B_n is the convex hull of finitely many doubly stochastic matrices. Therefore we only need to show that a doubly stochastic matrix $P \in B_n$ is an extreme point if and only if it is a permutation matrix. It follows from the definition that a permutation matrix is an extreme point. If $P = (p_{ij}) \in B_n$ is extreme, then the rank of the matrix with binding constraints must be n^2 by Proposition 4.5. The inequalities (4.30) and (4.31) are already binding, but they are linearly dependent (why?). Therefore we must have at least $n^2 - 2n + 1 = (n-1)^2$ binding inequalities among (4.29) so that P can have at most $2n - 1$ non-zero entries i.e., P has a row r with only one non-zero entry p_{rs}: 1 is the unique non-zero entry in row r and column s. Removing row r and column

---

[10]Garrett Birkhoff (1911–1996). American mathematician.

s you are left with an $(n-1) \times (n-1)$ doubly stochastic matrix and the result follows by induction. $\square$

4.8.1   *Perfect pairings and doubly stochastic matrices*

An elementary approach to doubly stochastic matrices and permutation matrices has been given by Knuth[11] in the beautiful exposition [Knuth (1981)].

An inductive procedure for writing a doubly stochastic matrix as a convex linear combination may be illustrated with the matrix

$$A = \begin{pmatrix} 0.4 & 0.2 & 0.4 \\ 0.2 & 0.5 & 0.3 \\ 0.4 & 0.3 & 0.3 \end{pmatrix}$$

from Example 4.25. As a first step we are looking for a permutation matrix P and a maximal $\lambda > 0$ such that $A - \lambda P \geq 0$. There are many choices for P. Choosing

$$P = \begin{pmatrix} 0 & 0 & 1 \\ 0 & 1 & 0 \\ 1 & 0 & 0 \end{pmatrix}$$

the maximal $\lambda > 0$ is $\lambda = 0.4$ and

$$A - \lambda P = \begin{pmatrix} 0.4 & 0.2 & 0.0 \\ 0.2 & 0.1 & 0.3 \\ 0.0 & 0.3 & 0.3 \end{pmatrix}.$$

Continuing this procedure we get

$$\begin{pmatrix} 0.4 & 0.2 & 0.0 \\ 0.2 & 0.1 & 0.3 \\ 0.0 & 0.3 & 0.3 \end{pmatrix} - 0.3 \begin{pmatrix} 1 & 0 & 0 \\ 0 & 0 & 1 \\ 0 & 1 & 0 \end{pmatrix} = \begin{pmatrix} 0.1 & 0.2 & 0.0 \\ 0.2 & 0.1 & 0.0 \\ 0.0 & 0.0 & 0.3 \end{pmatrix}$$

$$\begin{pmatrix} 0.1 & 0.2 & 0.0 \\ 0.2 & 0.1 & 0.0 \\ 0.0 & 0.0 & 0.3 \end{pmatrix} - 0.2 \begin{pmatrix} 0 & 1 & 0 \\ 1 & 0 & 0 \\ 0 & 0 & 1 \end{pmatrix} = \begin{pmatrix} 0.1 & 0.0 & 0.0 \\ 0.0 & 0.1 & 0.0 \\ 0.0 & 0.0 & 0.1 \end{pmatrix}$$

(4.32)

and finally

$$\begin{pmatrix} 0.1 & 0.0 & 0.0 \\ 0.0 & 0.1 & 0.0 \\ 0.0 & 0.0 & 0.1 \end{pmatrix} - 0.1 \begin{pmatrix} 1 & 0 & 0 \\ 0 & 1 & 0 \\ 0 & 0 & 1 \end{pmatrix} = \begin{pmatrix} 0.0 & 0.0 & 0.0 \\ 0.0 & 0.0 & 0.0 \\ 0.0 & 0.0 & 0.0 \end{pmatrix}.$$

---

[11]Donald E. Knuth (1938–). American mathematician and computer scientist.

By back substitution we now recover the convex combination in Example 4.25.

To make sure that this procedure works (termination of the procedure follows, since at least one positive entry turns into zero in each step) it suffices to prove that given any doubly stochastic matrix $A = (a_{ij})$ (or more precisely μA, where $0 < \mu \leq 1$), there exists a permutation matrix P and $\lambda > 0$, such that $A - \lambda P \geq 0$. This is not a problem when every entry in A is positive. When some entries are 0, a combinatorial tool is invoked: the existence of P with $\lambda > 0$ means that there exists a bijective map $\sigma : \{1, \ldots, n\} \to \{1, \ldots, n\}$ (a permutation), such that $a_{i,\sigma(i)} \neq 0$. Such a map is called a perfect pairing between rows and columns.

An example of a perfect pairing consists of the entries $a_{11}, a_{23}, a_{34}, a_{42}$ in

$$A = \begin{pmatrix} 0.5 & 0.5 & 0 & 0 \\ 0.5 & 0 & 0.3 & 0.2 \\ 0 & 0 & 0.7 & 0.3 \\ 0 & 0.5 & 0 & 0.5 \end{pmatrix}.$$

This corresponds to the permutation $\sigma(1) = 1$, $\sigma(2) = 3$, $\sigma(3) = 4$ and $\sigma(4) = 2$. Notice that a_{12}, a_{23} is not part of a perfect pairing. One needs to prove that every doubly stochastic matrix has a perfect pairing. This is done by a classical result from graph theory that can be phrased as follows.

Let M be a group of n men and W a group of n women engaging in "speed dating". It is known in advance that certain pairs of men and women have met before. Is it possible to pair all the men and women so that each pair have met before? Such a paring is called perfect. A *perfect pairing* is certainly impossible if there exists subsets $S \subseteq M$ and $T \subseteq W$ such that $|S| > |T|$ and men in S have only met women in T. The interesting result is the converse statement due to Hall.[12] The following proof of his result leads to a rather surprising algorithm for obtaining a perfect pairing if it exists (see Example 4.28).

Theorem 4.27 (Hall's marriage theorem, 1935). *If M and W do not have a perfect pairing, there exist subsets $S \subseteq M$ and $T \subseteq W$ with $|S| > |T|$, such that men in S have only met women in T.*

Proof. Suppose that k men is paired with k women, where k is chosen maximally. Then $k < n$. Let $M' \subsetneq M$ denote the men and $W' \subsetneq W$ the

---

[12]Philip Hall (1904–1982). English mathematician.

women in the pairing. Suppose that $m \notin M'$. Consider sequences

$$(m_1, w_1, m_2, w_2, m_3, \ldots, w_{r-1}, m_r, w_r)$$

with $m_1 = m$, $m_i \in M$ and $w_i \in W$, where m_i knows, but is not paired with w_i and w_i is paired with m_{i+1} for $i = 1, \ldots, r-1$. A crucial observation is that w_r in the end of the sequence must be paired. If this was not so, we could rearrange and pair w_r with m_r, w_{r-1} with m_{r-1}, $\ldots$, w_1 with m_1 thereby obtaining a pairing with more than k elements. Let S be the subset of men and T the subset of women appearing in such sequences. Then men in S do not know women outside T. Since m is not paired we must have $|S| > |T|$.   □

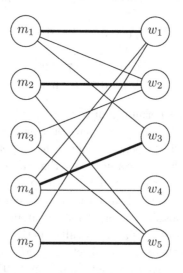

Figure 4.5: A speed dating scenario with 5 men and 5 women. The nodes on the left represent the men and the nodes on the right the women. The edges indicate if a pair has met before. A thick edge indicates that a pair is in the pairing.

Example 4.28. In Figure 4.5 m_1 is paired with w_1, m_2 with w_2, m_4 with w_3 and m_5 with w_5. This pairing is not perfect since m_3 is not paired. Starting with m_3 we build the sequence

$$t = (m_3, w_2, m_2, w_5, m_5, w_1, m_1, w_3, m_4, w_4)$$

using the procedure in the proof of Theorem 4.27. Here w_4 is not paired. We can therefore go backwards in t and

(1) insert the pair w_4 and m_4 and   remove the pair m_4 and w_3
(2) insert the pair w_3 and m_1 and   remove the pair m_1 and w_1
(3) insert the pair w_1 and m_5 and   remove the pair m_5 and w_5
(4) insert the pair w_5 and m_2 and   remove the pair m_2 and w_2
(5) insert the pair w_2 and m_3.

This gives the perfect pairing shown in Figure 4.6.

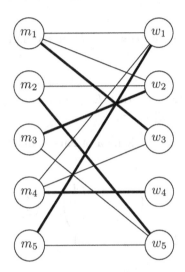

Figure 4.6: The perfect pairing in Example 4.28.

However if the edge from m_5 to w_1 is deleted no such trick is possible and we would end up with the subsets $S = \{m_3, m_2, m_5\} \subseteq M$ and $T = \{w_2, w_5\}$ alluded to in Theorem 4.27. In this case there is no perfect pairing.

Let us interpret doubly stochastic matrices in this framework. Let $A = (a_{ij})$ be an $n \times n$ doubly stochastic matrix. We define row i to have met column j if $a_{ij} > 0$. We wish to prove the existence of a perfect pairing of rows and columns in A.

If there is no such perfect pairing, Theorem 4.27 translates into the existence of a subset S of the rows and a subset T of the columns with $|S| > |T|$ such that $a_{ij} > 0$ and $i \in S$ implies $j \in T$ for $j = 1, \ldots, n$. Therefore

$$\sigma = \sum_{i \in S, j \in T} a_{ij} = |S|.$$

On the other hand σ is the sum of a subset of the entries in the columns in T. Therefore $\sigma \leq |T|$ contradicting $\sigma = |S| > |T|$.

4.9   Exercises

Exercise 4.1. Let $P = \{x \in \mathbb{R}^3 \mid Ax \leq b\}$ be a polyhedron in $\mathbb{R}^3$ and suppose that $z \notin P$. Give a criterion detecting if an extreme point $x \in \text{ext}(P)$ is visible from z.

Exercise 4.2. Compute the edges of P in Example 4.6 using Proposition 4.3.

Exercise 4.3. Let $Q = \text{conv}(\{v_1, \ldots, v_m\})$, where $v_i \in \mathbb{R}^d$. Suppose that $Q = \{x \in \mathbb{R}^d \mid Ax \leq b\}$. Prove that $E = \text{conv}(\{v_i, v_j\})$ is an edge of Q if $\text{rk } A_z = d - 2$, where $z = \frac{1}{2}(v_i + v_j)$.

Exercise 4.4. Show using the lemma of Farkas that

   (i) The equation

$$\begin{pmatrix} 1 & 2 & 3 \\ 3 & 1 & 5 \end{pmatrix} \begin{pmatrix} x \\ y \\ z \end{pmatrix} = \begin{pmatrix} 5 \\ 2 \end{pmatrix}$$

   is unsolvable with $x \geq 0, y \geq 0$ and $z \geq 0$.

  (ii) The equation

$$\begin{pmatrix} 1 & 2 & 3 & 1 \\ 3 & 1 & 5 & 1 \\ 1 & 2 & 1 & 1 \end{pmatrix} \begin{pmatrix} x \\ y \\ z \\ w \end{pmatrix} = \begin{pmatrix} 2 \\ 1 \\ 1 \end{pmatrix}$$

   is unsolvable with $x \geq 0, y \geq 0, z \geq 0$ and $w \geq 0$.

Exercise 4.5. Let $C \subseteq \mathbb{R}^n$ be a finitely generated cone.

   (i) Prove that if $x \notin C$, there exists $\alpha \in C^\circ$ with $\alpha^t x > 0$.

  (ii) Prove that

$$(C^\circ)^\circ = C.$$

Exercise 4.6. Let $P = \{x \in \mathbb{R}^d \mid Ax \leq b\} \subseteq \mathbb{R}^d$ be a polyhedron. Prove that if P contains a line $\{v + tu \mid t \in \mathbb{R}\}$ with $u \neq 0$, then $Au = 0$.

Exercise 4.7. Prove that the recession cone of a polyhedron $P = \{x \in \mathbb{R}^n \mid Ax \leq b\}$ is

$$\text{rec}(P) = \{x \in \mathbb{R}^n \mid Ax \leq 0\}.$$

Exercise 4.8. Use (4.13) to prove that a steady state for a Markov chain with

$$P = \begin{pmatrix} p_{11} & p_{21} \\ p_{12} & p_{22} \end{pmatrix}$$

and $p_{12} + p_{21} > 0$ is

$$x = \left(\frac{p_{21}}{p_{12} + p_{21}}, \frac{p_{12}}{p_{12} + p_{21}} \right).$$

Exercise 4.9. Show that a bounded polyhedron is a polytope. Is a polytope a bounded polyhedron?

Exercise 4.10. Give an example of a doubly stochastic 3×3 matrix with non-zero pairwise different entries. The procedure outlined in (4.32) stops in four steps. Can you give an example of a doubly stochastic 3×3 matrix, where it stops in five or six steps with suitable choices of permutation matrices along the way?

Exercise 4.11. Let $P_m \subseteq \mathbb{R}^{m \times m}$ be the polyhedron given by

$$x_{ij} \geq 0, \qquad \text{for } i, j = 1, \ldots, m$$
$$x_{i1} + \cdots + x_{im} \leq 1, \qquad \text{for } i = 1, \ldots, m$$
$$x_{1j} + \cdots + x_{mj} \leq 1, \qquad \text{for } j = 1, \ldots, m$$
$$\sum_{1 \leq i,j \leq m} x_{ij} \geq m - 1.$$

Prove that P_m is a polytope and that $\dim P_m = m^2$. Construct a natural affine map f from P_n to the Birkhoff polytope B_{n+1}. Use f to prove that the dimension of the Birkhoff polytope is

$$\dim B_n = (n-1)^2.$$

Exercise 4.12. Prove that a doubly stochastic matrix $A \in B_n$ is a convex combination of at most $n^2 - 2n + 2$ permutation matrices.

Exercise 4.13. Prove that the graph below (see Example 4.28) does not have a perfect pairing.

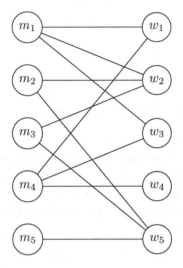

Chapter 5

Computations with polyhedra

The Minkowski-Weyl theorem (Theorem 4.23) says that a polyhedron $P \subseteq \mathbb{R}^d$ is representable as the Minkowski sum

$$P = C + Q = \{u + v \mid u \in C, v \in Q\}, \tag{5.1}$$

where C is a finitely generated cone and Q the convex hull of finitely many points in $\mathbb{R}^d$. We call this the *vertex representation* of P. The theorem also says that a subset P as in (5.1) is a polyhedron i.e., it is representable as an intersection

$$P = \{x \in \mathbb{R}^d \mid Ax \leq b\}$$

of half spaces. We call this the *half space representation* of P.

In particular, the convex hull of a finite set of points in $\mathbb{R}^d$ has a half space representation $\{x \in \mathbb{R}^d \mid Ax \leq b\}$. This representation makes it easy to decide if a given point is inside the convex hull. An effective computational procedure for finding A and b is extremely useful in many areas of pure and applied mathematics. On the other hand, the conversion from half space to vertex representation may be viewed as the equivalent of solving systems of linear equations in linear algebra. These conversions can both be reduced to the problem of finding a matrix R, such that (A, R) is a double description pair (see Definition 4.11), where A is a given matrix.

In this chapter we will introduce the *double description method*. This is an algorithm for computing R inductively from the rows of A. The double description method was originally introduced in the context of game theory in the rough procedure outlined in the brief §10 of [Motzkin *et al.* (1953)]. Our introduction is inspired by the lucid exposition in [Fukuda and Prodon (1996)].

The *simplex algorithm* is a procedure for finding an optimum for a linear function on a polyhedron. Heavily influenced by practical applications, it was proposed by George Dantzig in 1947, when he was working for the US Air Force. From a mathematical point of view it is not supposed to work effectively and may be downplayed as a computational hack not even guaranteed to terminate. In fact, John von Neumann "always contended with Dantzig that the simplex method would take an absurdly long amount of time to solve linear programming problems".[1]

The simplex algorithm has, however, been observed to function beautifully[2] and continues to serve as a stellar example of the power of exploring mathematics using a computer. Emphasizing the simplex tableau for general polyhedra, we give an introduction to the simplex algorithm at the end of this chapter.

5.1   Extreme rays and minimal generators in convex cones

Intuitively the polyhedral cone $\{x \in \mathbb{R}^d \mid Ax \leq 0\}$ is minimally generated by rays given by intersections of supporting half spaces.

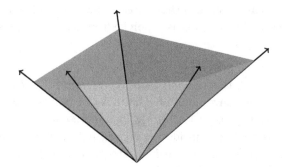

Figure 5.1: A polyhedral cone in $\mathbb{R}^3$ cut out by 5 half spaces and generated by 5 extreme rays.

We make this precise in the context of general convex cones. First recall the definition of a face of a convex subset (Definition 3.7).

Definition 5.1. Let C be a convex cone. A non-zero $r \in C$ is called an *extreme ray* in C if $\text{cone}(\{r\})$ is a face of C.

---

[1]From an interview with Philip Wolfe, November 28, 1972. Archives Center, National Museum of American History. Wolfe also says: "It appears to be, oh, so far as I know, the one place where Johnny went very badly wrong."

[2]A mathematical explanation for the observed efficiency of the simplex method has been given recently in [Spielman and Teng (2001)].

One can check that a non-zero vector $r \in C$ is an extreme ray if and only if

$$r = x + y \quad \text{implies} \quad x, y \in \text{cone}(\{r\}) \tag{5.2}$$

for $x, y \in C$. By iterating (5.2) we get the following result, which shows that extreme rays always appear in a generating set for a convex cone.

Lemma 5.2. *Let $C = \text{cone}(\{r_1, \ldots, r_n\}) \subseteq \mathbb{R}^d$ be a finitely generated cone and $r \in C$ an extreme ray. Then r is a positive multiple of some r_j for $j = 1, \ldots, n$.*

Because of Lemma 5.2 we will also refer to an extreme ray in a convex cone as a *minimal generator*.

5.2 Minimal generators of a polyhedral cone

We will study extreme rays in greater detail for polyhedral cones. Since a polyhedron has only finitely many faces, we know that a polyhedral cone can only have finitely many extreme rays up to multiplication by a positive scalar.

We will prove that a pointed polyhedral cone is generated by its extreme rays. Using the notation introduced in §4.1 this is done below (see Theorem B.9 for the relevant linear algebra).

Theorem 5.3. *Let A be a matrix of full rank d with rows $a_1, \ldots, a_m \in \mathbb{R}^d$. Consider the polyhedral cone*

$$C = \{x \in \mathbb{R}^d \mid Ax \leq 0\}.$$

(1) *A vector $r \in C$ is an extreme ray if and only if $\text{rk } A_r = d - 1$.*
(2) *Two extreme rays $r_1, r_2 \in C$ span a two-dimensional face $\text{cone}(\{r_1, r_2\})$ of C if and only if $\text{rk } A_J = d - 2$, where $J = I(r_1) \cap I(r_2)$.*
(3) *Every element of C is a sum of (finitely many) extreme rays in C.*

Proof. The proof of (1) follows from Proposition 4.3 using that A has full rank.

For the proof of (2), assume first that $r_1, r_2 \in C$ are extreme rays, such that $F = \text{cone}(\{r_1, r_2\}) \subseteq C$ is a face of dimension 2 of C. Then $F = \{x \in C \mid A_J x = 0\}$, where $\text{rk } A_J = d - 2$ and $J = I(r)$ for $r \in F$ by Proposition 4.3. If $r = \lambda r_1 + \mu r_2$, then $\lambda, \mu > 0$ and $J = I(r_1) \cap I(r_2)$. If on the other hand, $r_1, r_2 \in C$ are extreme rays with $\text{rk } A_J = d - 2$, where

$J = I(r_1) \cap I(r_2)$, then r_1, r_2 form a basis of the two-dimensional vector space $\{x \in \mathbb{R}^d \mid A_J x = 0\}$. Clearly, $F = \{x \in C \mid A_J x = 0\}$ is a face of C containing cone($\{r_1, r_2\}$). If $x \in F$, then $x = \lambda r_1 + \mu r_2$. We may find row vectors a_i, a_j of A with $a_i r_1 < 0, a_i r_2 = 0$ and $a_j r_1 = 0, a_j r_2 < 0$. This proves that $\lambda, \mu \geq 0$.

We finish by proving (3) providing an algorithm for writing a given $z \in C$ as a finite sum of extreme rays. If $\mathrm{rk}\, A_z = d$, then $z = 0$ and z is the empty sum $(= 0)$ of extreme rays. If $\mathrm{rk}\, A_z = d - 1$, z is an extreme ray by (1) and we are done. If $\mathrm{rk}\, A_z \leq d - 2$ we can find $i, j \notin I(z)$, such that $\mathrm{rk}\, A_{\{i,j\} \cup I(z)} = \mathrm{rk}\, A_z + 2$. The crucial point is the existence of a vector $v \in \mathbb{R}^d$, such that $A_z v = 0$, $a_i v > 0$ and $a_j v < 0$. Such a vector can be constructed putting $v = v_1 + v_2$, where $v_1, v_2 \in \mathbb{R}^d$ satisfy $A_z v_1 = A_z v_2 = 0$, $a_i v_1 > 0, a_j v_1 = 0$ and $a_i v_2 = 0, a_j v_2 < 0$.

In general we have the identity

$$z = \frac{\mu}{\lambda + \mu}(z - \lambda v) + \frac{\lambda}{\lambda + \mu}(z + \mu v)$$

for every $\lambda, \mu > 0$. With

$$\lambda = \lambda^* := \sup\{\lambda > 0 \mid z - \lambda v \in C\} = \min\left\{ \frac{a_k z}{a_k v} \,\middle|\, a_k v < 0, k \notin I(z) \right\}$$

$$\mu = \mu^* := \sup\{\mu > 0 \mid z + \mu v \in C\} = \min\left\{ -\frac{a_k z}{a_k v} \,\middle|\, a_k v > 0, k \notin I(z) \right\},$$

it follows that $x_1 = z - \lambda^* v, x_2 = z + \mu^* v \in C$ are both non-zero, $\mathrm{rk}\, A_{x_1} > \mathrm{rk}\, A_z$ and $\mathrm{rk}\, A_{x_2} > \mathrm{rk}\, A_z$. This procedure can be repeated if necessary with x_1 or x_2. $\qquad\square$

The proof of (3) in Theorem 5.3 contains a very explicit algorithm for writing $x \in C$ as a sum of minimal generators or extreme rays. Here is an example illustrating the mechanics of this procedure using the notation in the proof.

Example 5.4. Let

$$A = \begin{pmatrix} -1 & 0 & 0 \\ 0 & -1 & 0 \\ 0 & 0 & -1 \\ 1 & -1 & -1 \\ 1 & -2 & 1 \end{pmatrix}.$$

Then $z = (1, 1, 1) \in C = \{x \in \mathbb{R}^3 \mid Ax \leq 0\}$, $I(z) = \{5\}$ and $A_z = (1, -2, 1)$. Therefore $\mathrm{rk}\, A_z = 1 < 2 = 3 - 1$ and z is not an extreme ray. Following the general method in the proof we put $i = 1$, $j = 2$ and $v_1 = (-1, 0, 1)$, $v_2 = (0, 1, 2)$. With $v = v_1 + v_2 = (-1, 1, 3)$ we have

$$a_1 v = 1, \qquad a_2 v = -1, \qquad a_3 v = -3, \qquad a_4 v = -5$$

and $a_1 z = a_2 z = a_3 z = a_4 z = -1$. Therefore $\lambda^* = \frac{1}{5}$ and $\mu^* = 1$ and

$$x_1 = (1, 1, 1) - \tfrac{1}{5}(-1, 1, 3) = (\tfrac{6}{5}, \tfrac{4}{5}, \tfrac{2}{5})$$
$$x_2 = (1, 1, 1) + (-1, 1, 3) \quad = (0, 2, 4)$$

with $z = \frac{5}{6}x_1 + \frac{1}{6}x_2$. You may want to scale x_1 and x_2 by the positive scalars $\frac{5}{2}$ and $\frac{1}{2}$ respectively to work with the more aesthetically pleasing vectors $(3, 2, 1)$ and $(0, 1, 2)$. From a mathematical point of view this is unimportant. Finally $I(x_1) = \{4, 5\}$ and $I(x_2) = \{1, 5\}$ and you can check that both x_1 and x_2 are extreme rays.

Let $C = \{x \in \mathbb{R}^d \mid Ax \leq 0\}$, where A is a matrix of full rank d. Theorem 5.3(1) gives a procedure for finding the minimal generators (extreme rays) $r_1, \ldots, r_N \in C$ with

$$C = \text{cone}(\{r_1, \ldots, r_N\}).$$

Every extreme ray $r \in C$ satisfies $\text{rk}\, A_{I(r)} = d - 1$. With this in mind we can find the extreme rays by running through linear independent subsets $J \subseteq [m]$ of the rows with $d - 1$ elements writing $\{x \in \mathbb{R}^d \mid A_J x = 0\} = \mathbb{R}z$. The subset J defines an extreme ray in C if $z \in C$ or $-z \in C$. Here is an example of how this works.

Example 5.5. Consider the polyhedral cone $C = \{x \in \mathbb{R}^3 \mid Ax \leq 0\}$ defined in Example 5.4. To find the extreme rays of C based on Theorem 5.3(1), one must potentially run through the $\binom{5}{2} = 10$ possible subsets J of two rows of A, verify that A_J has rank 2, find z with $\{x \in \mathbb{R}^3 \mid A_J x = 0\} = \mathbb{R}z$ and then check if $z \in C$ or $-z \in C$. For $J = \{3, 4\}$ we have

$$A_J = \begin{pmatrix} 0 & 0 & -1 \\ 1 & -1 & -1 \end{pmatrix}$$

and $\{x \in \mathbb{R}^3 \mid A_J x = 0\} = \mathbb{R}z$ with $z = (1, 1, 0)$. Since $Az \leq 0$ it follows that $z \in C$ and that z is an extreme ray of C. If $J = \{4, 5\}$,

$$A_J = \begin{pmatrix} 1 & -1 & -1 \\ 1 & -2 & 1 \end{pmatrix}$$

and $\{x \in \mathbb{R}^3 \mid A_J x = 0\} = \mathbb{R}z$ for $z = (3,\ 2,\ 1)$ retrieving one of the extreme rays from Example 5.4. Finally if $J = \{2,5\}$,

$$A_J = \begin{pmatrix} 0 & -1 & 0 \\ 1 & -2 & 1 \end{pmatrix}$$

and $\{x \in \mathbb{R}^3 \mid A_J x = 0\} = \mathbb{R}z$ for $z = (-1,\ 0,\ 1)$. Since neither $z \in C$ nor $-z \in C$, the subset $J = \{2,5\}$ does not give an extreme ray of C.

5.3   The double description method

The *double description method* is an algorithm for computing a matrix R in a double description pair (A, R) given the matrix A (see Definition 4.11).

The fundamental step lies in the computation of the extreme points in the intersection $C \cap H^-$ of a convex hull $C = \mathrm{conv}(\{v_1, \ldots, v_m\})$ with an affine half space H^-. Geometrically this operation is rather simple and can be understood from a sketch intersecting a polygon with a half plane (see Figure 5.2).

Figure 5.2: The basic step of the double description method: The extreme points of $C \cap H^-$ can be computed intersecting H with lines between the extreme points of C on both sides of H.

In the context of the double description method, an important consequence of Theorem 5.3(1) is the following proposition, which immediately gives R in (A, R) when A is invertible.

Proposition 5.6. *Let A be an invertible $d \times d$ matrix. Then*

$$C = \{x \in \mathbb{R}^d \mid Ax \leq 0\} = \{Rz \mid z \geq 0\},$$

where $R = -A^{-1}$. The extreme rays of C are the columns of R.

Proof. This is left to the reader (see Exercise 5.2). □

The following result, which is also a consequence of Theorem 5.3(1), is essentially the content of the wording in §10 of [Motzkin *et al.* (1953)].

Proposition 5.7. *Let* $\alpha = (\alpha_1, \ldots, \alpha_d) \in \mathbb{R}^d$. *Then the minimal generators of* $C = \{x \in \mathbb{R}^d \mid Ax \le 0\}$, *where* A *is the* $(d+1) \times d$ *matrix*

$$A = \begin{pmatrix} \alpha_1 & \alpha_2 & \cdots & \alpha_d \\ -1 & 0 & \cdots & 0 \\ 0 & -1 & \cdots & 0 \\ \vdots & \vdots & \ddots & \vdots \\ 0 & 0 & \cdots & -1 \end{pmatrix}$$

are

$$\{e_i \mid \alpha_i \le 0\} \cup \{\alpha_i e_j - \alpha_j e_i \mid \alpha_i > 0, \alpha_j < 0\},$$

where $e_1, \ldots, e_d \in \mathbb{R}^d$ *are the canonical basis vectors.*

Proof. This is an application of Theorem 5.3(1). If I is a subset of the rows of A, such that $\operatorname{rk} A_I = d - 1$, then $\{v \in \mathbb{R}^d \mid A_I x = 0\} = \mathbb{R}z$ for some $z \in \mathbb{R}^d$. If I gives rise to a minimal generator, then $z \in C$ or $-z \in C$. If the first row is not in I, then $z = e_j$ for some j and therefore $\alpha_j \le 0$. If the first row is in I, then $z = \lambda e_i + \mu e_j$ for $\lambda, \mu \ge 0$. If $\lambda, \mu > 0$ we must have $\lambda \alpha_i + \mu \alpha_j = 0$. Therefore we may assume $\alpha_i > 0, \alpha_j < 0$ and z is a positive multiple of $\alpha_i e_j - \alpha_j e_i$. □

An important consequence of this result is the fundamental step in the double description method (compare this with Figure 5.2).

Corollary 5.8. *Let* $r_1, \ldots, r_m \in \mathbb{R}^d$. *If* $\alpha \in \mathbb{R}^d$, *then*

$$\operatorname{cone}(\{r_1, \ldots, r_m\}) \cap \{x \in \mathbb{R}^d \mid \alpha^t x \le 0\}$$
$$= \operatorname{cone}\left(\{r_j \mid \alpha^t r_j \le 0\} \cup \{(\alpha^t r_i)r_j - (\alpha^t r_j)r_i \mid \alpha^t r_i > 0, \ \alpha^t r_j < 0\}\right).$$
$$(5.3)$$

Proof. Let R denote the $d \times m$ matrix with columns $r_1, \ldots, r_m$. Then $\operatorname{cone}(\{r_1, \ldots, r_m\}) = \{Rz \mid z \ge 0\}$ and

$$\{Rz \mid z \ge 0\} \cap \{x \in \mathbb{R}^d \mid \alpha^t x \le 0\}$$
$$= \{Rz \mid (\alpha^t R)z \le 0, \ z \ge 0\}.$$

Now the result follows from Proposition 5.7. □

Corollary 5.8 gives a powerful inductive method for computing generators for a polyhedral cone.

Example 5.9. Let

$$C = \left\{ v \in \mathbb{R}^3 \,\middle|\, \begin{pmatrix} -1 & 0 & 0 \\ 0 & -1 & 0 \\ 0 & 0 & -1 \\ 1 & -1 & -1 \end{pmatrix} v \le 0 \right\}.$$

Let A be the 3×3 matrix consisting of the first three rows of the matrix defining C. Then

$$C' = \{x \in \mathbb{R}^3 \mid Ax \le 0\} = \mathrm{cone}(\{(1,\,0,\,0),(0,\,1,\,0),(0,\,0,\,1)\})$$

by Proposition 5.6. With $\alpha = (1,\,-1,\,-1)$ we get

$$
\begin{aligned}
C &= C' \cap \{x \in \mathbb{R}^3 \mid \alpha^t x \le 0\} \\
&= \mathrm{cone}\left(\{(0,\,1,\,0),(0,\,0,\,1),(1,\,1,\,0),(1,\,0,\,1)\}\right),
\end{aligned}
$$

using that $\alpha^t(1,\,0,\,0) = 1$, $\alpha^t(0,\,1,\,0) = -1$, $\alpha^t(0,\,0,\,1) = -1$ in Corollary 5.8.

Using Theorem 5.3(1) you can verify that these generators of C are extreme rays and therefore form a minimal generating set. Now let $\alpha = (1,\,-2,\,1)$. Since $\alpha^t(0,\,1,\,0) = -2$, $\alpha^t(0,\,0,\,1) = 1$, $\alpha^t(1,\,1,\,0) = -1$, $\alpha^t(1,\,0,\,1) = 2$, Corollary 5.8 gives

$$\left\{ v \in \mathbb{R}^3 \,\middle|\, \begin{pmatrix} -1 & 0 & 0 \\ 0 & -1 & 0 \\ 0 & 0 & -1 \\ 1 & -1 & -1 \\ 1 & -2 & 1 \end{pmatrix} v \le 0 \right\} = C \cap \{x \in \mathbb{R}^3 \mid \alpha^t x \le 0\}$$

$$= \mathrm{cone}(\{(0,\,1,\,0),(1,\,1,\,0),(1,\,1,\,1),(3,\,2,\,1),(0,\,1,\,2)\}).$$

Here you can check that $(1,\,1,\,1)^t$ is not an extreme ray. It is therefore a redundant generator in agreement with the computation in Example 5.4, which showed that

$$(1,\,1,\,1) = \tfrac{1}{3}(3,\,2,\,1) + \tfrac{1}{3}(0,\,1,\,2).$$

The double description method in the incarnation of Corollary 5.8 is doomed. It simply produces too many redundant generators in each step. A remedy is the following nice result, which says that $(\alpha^t r_i)r_j - (\alpha^t r_j)r_i$ is only needed in Corollary 5.8, when $\mathrm{cone}(\{r_i, r_j\})$ is a face of C. This can be detected by linear algebra using Theorem 5.3(2).

Corollary 5.10. *Suppose that $r_1, \ldots, r_m \in \mathbb{R}^d$ are the extreme rays of*

$$C = \mathrm{cone}(\{r_1, \ldots, r_m\}) = \{x \in \mathbb{R}^d \mid Ax \leq 0\}$$

for some $n \times d$ matrix A. Let $\alpha \in \mathbb{R}^d$. Then

$$\{r_j \mid \alpha^t r_j \leq 0\} \cup \{(\alpha^t r_i) r_j - (\alpha^t r_j) r_i$$
$$\mid \alpha^t r_i > 0,\ \alpha^t r_j < 0,\ and\ \mathrm{cone}(\{r_i, r_j\})\ is\ a\ face\ of\ C\}$$

are the extreme rays of $C \cap \{x \in \mathbb{R}^d \mid \alpha^t x \leq 0\}$.

Proof. Notice first that Corollary 5.8 implies that the extreme rays of $C \cap \{x \in \mathbb{R}^d \mid \alpha^t x \leq 0\}$ are among

$$\{r_j \mid \alpha^t r_j \leq 0\} \cup \{(\alpha^t r_i) r_j - (\alpha^t r_j) r_i \mid \alpha^t r_i > 0,\ \alpha^t r_j < 0\}.$$

The extreme rays among these are as claimed by using Theorem 5.3(2) observing that

$$I\big((\alpha^t r_i) r_j - (\alpha^t r_j) r_i\big) = \big(I(r_j) \cap I(r_i)\big) \cup \{n+1\},$$

where $\alpha^t r_i > 0, \alpha^t r_j < 0$ and the index $n+1$ refers to the last row of the matrix

$$\begin{pmatrix} A \\ \alpha^t \end{pmatrix}. \qquad \square$$

Example 5.11. Revisiting Example 5.9, the redundant generator $(1, 1, 1)$ came from Corollary 5.8 as a result of the computation

$$(\alpha^t(1, 0, 1))(0, 1, 0) + (\alpha^t(0, 1, 0))(1, 0, 1)$$
$$= 2(0, 1, 0) + 2(1, 0, 1)$$
$$= (2, 2, 2).$$

Corollary 5.10 reveals in advance that this is a redundant computation, since $\mathrm{cone}(\{(1, 0, 1), (0, 1, 0)\})$ is not a face of

$$C = \left\{ v \in \mathbb{R}^3 \ \middle| \ \begin{pmatrix} -1 & 0 & 0 \\ 0 & -1 & 0 \\ 0 & 0 & -1 \\ 1 & -1 & -1 \end{pmatrix} v \leq 0 \right\}.$$

This follows by Theorem 5.3(2), since $I_1 = I((1, 0, 1)) = \{2, 4\}$, $I_2 = I((0, 1, 0)) = \{1, 3\}$ and $I_1 \cap I_2 = \emptyset$.

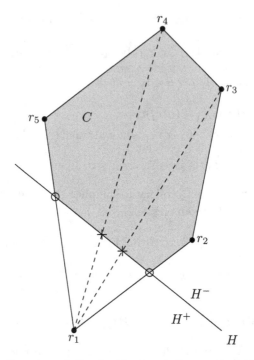

Figure 5.3: One step of the double description method illustrating the computation of $C \cap H^-$.

Figure 5.3 shows a picture illustrating the mechanics of and differences between Corollary 5.8 and Corollary 5.10. Here C is a cone in $\mathbb{R}^3$ generated by the (extreme) vectors $r_1, \ldots, r_5$ (viewed as a two-dimensional slice), $H = \{x \in \mathbb{R}^3 \mid \alpha^t x = 0\}$, $H^- = \{x \in \mathbb{R}^3 \mid \alpha^t x < 0\}$ and $H^+ = \{x \in \mathbb{R}^3 \mid \alpha^t x > 0\}$. The extreme ray r_1 in $C \cap H^+$ is part of two two-dimensional faces in C neighboring the two extreme rays r_2 and r_5 in $C \cap H^-$. Corollary 5.10 only adds the two new extreme rays

$$(\alpha^t r_1) r_2 - (\alpha^t r_2) r_1$$
$$(\alpha^t r_1) r_5 - (\alpha^t r_5) r_1$$

marked o in $C \cap H^-$. The dashed lines indicate the computation of the redundant generators

$$(\alpha^t r_1) r_4 - (\alpha^t r_4) r_1$$
$$(\alpha^t r_1) r_3 - (\alpha^t r_3) r_1$$

marked + in $C \cap H^-$ present in the procedure of Corollary 5.8.

5.3.1 *Converting from half space to vertex representation*

Suppose that $P = \{x \in \mathbb{R}^d \mid Ax \leq b\}$ is a polyhedron. As in the proof of Theorem 4.23 we define the polyhedral cone

$$\hat{P} = \left\{ \begin{pmatrix} x \\ z \end{pmatrix} \middle| x \in \mathbb{R}^d,\ z \in \mathbb{R},\ Ax \leq bz,\ z \geq 0 \right\}.$$

Recall that

$$x \in P \iff \begin{pmatrix} x \\ 1 \end{pmatrix} \in \hat{P}. \tag{5.4}$$

In matrix notation, $\hat{P}$ is given by

$$\left\{ \begin{pmatrix} x \\ z \end{pmatrix} \in \mathbb{R}^{d+1} \middle| \begin{pmatrix} A & -b \\ 0 & -1 \end{pmatrix} \begin{pmatrix} x \\ z \end{pmatrix} \leq 0 \right\}.$$

Using the double description method we may compute vectors $u_1, \ldots, u_r$, $v_1, \ldots, v_s \in \mathbb{R}^d$, such that

$$\hat{P} = \operatorname{cone}\left(\left\{ \begin{pmatrix} u_1 \\ 0 \end{pmatrix}, \ldots, \begin{pmatrix} u_r \\ 0 \end{pmatrix}, \begin{pmatrix} v_1 \\ 1 \end{pmatrix}, \ldots, \begin{pmatrix} v_s \\ 1 \end{pmatrix} \right\} \right).$$

Here we may have $s = 0$ indicating that there are no v's above. This means that $P = \emptyset$ by (5.4). Assuming that $P \neq \emptyset$ it follows that

$$P = C + Q,$$

where $C = \operatorname{cone}(\{u_1, \ldots, u_r\})$ and $Q = \operatorname{conv}(\{v_1, \ldots, v_s\})$. This shows how to use the double description method in writing P as the Minkowski sum of a finitely generated cone and a polytope.

Example 5.12. Consider the set P of $(x, y, z) \in \mathbb{R}^3$ satisfying

$$
\begin{array}{rcrcrcl}
x & & & & & \geq & 1 \\
x & & & + & z & \geq & 2 \\
& & y & & & \geq & 1 \\
& & y & + & z & \geq & 2 \\
x & + & y & - & z & \leq & 4 \\
x & + & y & & & \leq & 5 .
\end{array}
$$

This is a polyhedron in half space representation. It is not clear in advance if P is bounded or even if $P \neq \emptyset$. To analyze P further we convert to vertex representation by considering the polyhedral cone

$$\hat{P} = \left\{ \begin{pmatrix} x \\ y \\ z \\ w \end{pmatrix} \in \mathbb{R}^4 \middle| \begin{pmatrix} -1 & 0 & 0 & 1 \\ -1 & 0 & -1 & 2 \\ 0 & -1 & 0 & 1 \\ 0 & -1 & -1 & 2 \\ 1 & 1 & -1 & -4 \\ 1 & 1 & 0 & -5 \\ 0 & 0 & 0 & -1 \end{pmatrix} \begin{pmatrix} x \\ y \\ z \\ w \end{pmatrix} \leq \begin{pmatrix} 0 \\ 0 \\ 0 \\ 0 \\ 0 \\ 0 \\ 0 \end{pmatrix} \right\}$$

in $\mathbb{R}^4$. The double description method applies and gives

$$\hat{P} = \text{cone} \left(\left\{ \begin{pmatrix} 0 \\ 0 \\ 1 \\ 0 \end{pmatrix}, \begin{pmatrix} 1 \\ 1 \\ 1 \\ 1 \end{pmatrix}, \begin{pmatrix} 1 \\ 4 \\ 1 \\ 1 \end{pmatrix}, \begin{pmatrix} 2 \\ 2 \\ 0 \\ 1 \end{pmatrix}, \begin{pmatrix} 4 \\ 1 \\ 1 \\ 1 \end{pmatrix} \right\} \right), \tag{5.5}$$

where all generators are extreme rays of $\hat{P}$. Therefore

$$P = C + Q$$

with

$$C = \text{cone} \left(\left\{ \begin{pmatrix} 0 \\ 0 \\ 1 \end{pmatrix} \right\} \right) \quad \text{and} \quad Q = \text{conv} \left(\left\{ \begin{pmatrix} 1 \\ 1 \\ 1 \end{pmatrix}, \begin{pmatrix} 1 \\ 4 \\ 1 \end{pmatrix}, \begin{pmatrix} 2 \\ 2 \\ 0 \end{pmatrix}, \begin{pmatrix} 4 \\ 1 \\ 1 \end{pmatrix} \right\} \right).$$

5.3.2   *Converting from vertex to half space representation*

Suppose that

$$P = C + Q \subseteq \mathbb{R}^d$$

with

$$C = \text{cone}(\{u_1, \ldots, u_r\}) \quad \text{and} \quad Q = \text{conv}(\{v_1, \ldots, v_s\}).$$

To compute A and b with $\{x \in \mathbb{R}^d \mid Ax \le b\} = C + Q$ we consider the $(d+1) \times (r+s)$ matrix

$$R = \begin{pmatrix} u_1 & \cdots & u_r & v_1 & \cdots & v_s \\ 0 & \cdots & 0 & 1 & \cdots & 1 \end{pmatrix} \tag{5.6}$$

and notice that

$$x \in C + Q \quad \Longleftrightarrow \quad \begin{pmatrix} x \\ 1 \end{pmatrix} \in \{Rz \mid z \ge 0\}.$$

If $(\hat{A}, R)$ is a double description pair, then

$$C + Q = \left\{ x \in \mathbb{R}^d \,\middle|\, \hat{A} \begin{pmatrix} x \\ 1 \end{pmatrix} \le 0 \right\} = \{x \in \mathbb{R}^d \mid Ax \le b\},$$

where b is the negative of the last column and A the matrix consisting of the first d columns of $\hat{A}$. Observing that $(\hat{A}, R)$ is a double description pair if and only if $(R^t, \hat{A}^t)$ is a double description pair (see Lemma 4.21), we apply the double description method to the polyhedral cone

$$\{x \in \mathbb{R}^d \mid R^t x \le 0\}$$

to find $\hat{A}$.

Example 5.13. Write down a minimal set of inequalities in x, y and z describing if

$$\begin{pmatrix} x \\ y \\ z \end{pmatrix}$$

is in the convex hull of the vectors

$$\begin{pmatrix} 1 \\ 1 \\ 1 \end{pmatrix}, \quad \begin{pmatrix} 2 \\ 1 \\ 1 \end{pmatrix}, \quad \begin{pmatrix} 1 \\ 2 \\ 1 \end{pmatrix}, \quad \begin{pmatrix} 2 \\ 2 \\ 3 \end{pmatrix} \quad \text{and} \quad \begin{pmatrix} 1 \\ 1 \\ 5 \end{pmatrix}$$

in $\mathbb{R}^3$. This is solved as described above, first building the matrix

$$R = \begin{pmatrix} 1 & 2 & 1 & 2 & 1 \\ 1 & 1 & 2 & 2 & 1 \\ 1 & 1 & 1 & 3 & 5 \\ 1 & 1 & 1 & 1 & 1 \end{pmatrix}$$

as in (5.6). The double description method applied to $\{x \in \mathbb{R}^4 \mid R^t x \le 0\}$ gives

$$\hat{P} := \left\{ \begin{pmatrix} x_1 \\ x_2 \\ x_3 \\ x_4 \end{pmatrix} \middle| \begin{pmatrix} 1 & 1 & 1 & 1 \\ 2 & 1 & 1 & 1 \\ 1 & 2 & 1 & 1 \\ 2 & 2 & 3 & 1 \\ 1 & 1 & 5 & 1 \end{pmatrix} \begin{pmatrix} x_1 \\ x_2 \\ x_3 \\ x_4 \end{pmatrix} \le \begin{pmatrix} 0 \\ 0 \\ 0 \\ 0 \\ 0 \end{pmatrix} \right\} \tag{5.7}$$

$$= \text{cone} \left(\left\{ \begin{pmatrix} -2 \\ 4 \\ 1 \\ -7 \end{pmatrix}, \begin{pmatrix} -1 \\ 0 \\ 0 \\ 1 \end{pmatrix}, \begin{pmatrix} 0 \\ -1 \\ 0 \\ 1 \end{pmatrix}, \begin{pmatrix} 0 \\ 0 \\ -1 \\ 1 \end{pmatrix}, \begin{pmatrix} 2 \\ 2 \\ -1 \\ -5 \end{pmatrix}, \begin{pmatrix} 4 \\ -2 \\ 1 \\ -7 \end{pmatrix} \right\} \right) \tag{5.8}$$

$$= \{\hat{A}^t z \mid z \ge 0\}.$$

Here

$$\hat{A} = \begin{pmatrix} -2 & 4 & 1 & -7 \\ -1 & 0 & 0 & 1 \\ 0 & -1 & 0 & 1 \\ 0 & 0 & -1 & 1 \\ 2 & 2 & -1 & -5 \\ 4 & -2 & 1 & -7 \end{pmatrix}.$$

Therefore the inequalities

$$\begin{aligned}
-2x + 4y + z &\le 7 \\
-x \quad\quad\quad &\le -1 \\
-y \quad\quad &\le -1 \\
-z &\le -1 \\
2x + 2y - z &\le 5 \\
4x - 2y + z &\le 7
\end{aligned}$$

suffice to define the convex hull. The rows in $\hat{A}$ are all extreme rays in $\hat{P}$. Therefore none of the inequalities above are redundant in the sense that an inequality can be expressed as a conic combination of other inequalities.

5.3.3   Computing the convex hull

We will go deeper into the computation of the convex hull of a finite set of points based on the procedure outlined in §5.3.2. Suppose that we are given a set $V = \{v_1, \ldots, v_m\} \subseteq \mathbb{R}^d$ of points. In computing $Q = \text{conv}(V)$ as a minimal intersection of half spaces $\{x \in \mathbb{R}^d \mid Ax \leq b\}$, we need to compute a matrix $\hat{A}$ with

$$C = \{x \in \mathbb{R}^{d+1} \mid R^t x \leq 0\} = \{\hat{A}^t z \mid z \geq 0\},$$

where

$$R^t = \begin{pmatrix} v_1 & 1 \\ \vdots & \vdots \\ v_m & 1 \end{pmatrix}.$$

We will use the notation

$$\hat{A}^t = \begin{pmatrix} a_1 & \cdots & a_r \\ -b_1 & \cdots & -b_r \end{pmatrix},$$

where $a_1, \ldots, a_r \in \mathbb{R}^d$ and $b_1, \ldots, b_r \in \mathbb{R}$. With this convention A is the $r \times d$ matrix with row vectors $a_1, \ldots, a_r$ and $b = (b_1, \ldots, b_r)^t$. Inductively we may assume that $v_1, \ldots, v_m$ are extreme points of Q and that the columns of $\hat{A}^t$ are extreme rays of C. The i-th column $\hat{A}_i$ of $\hat{A}^t$ defines the affine hyperplane

$$H_i = \{x \in \mathbb{R}^d \mid a_i^t x = b_i\}.$$

Let $H_i^{++} = \{x \in \mathbb{R}^d \mid a_i^t x > b_i\}$, $H_i^- = \{x \in \mathbb{R}^d \mid a_i^t x \leq b_i\}$ and $H_i^{--} = \{x \in \mathbb{R}^d \mid a_i^t x < b_i\}$. Then

$$Q = H_1^- \cap \cdots \cap H_r^-$$

and $Q \cap H_i$ is a facet of Q. Two columns $\hat{A}_i$ and $\hat{A}_j$ combine to form a two-dimensional face of C if and only if $Q \cap H_i \cap H_j$ is a face of Q. Updating $\hat{A}$ to match $\text{conv}(Q \cup \{u\})$ when adding a point $u \in \mathbb{R}^d$ to V translates geometrically into

- adding a new hyperplane for every pair (H_i, H_j), where $Q \cap H_i \cap H_j$ a face of Q, $u \in H_i^{++}$ and $u \in H_j^{--}$

– deleting hyperplanes H_k with $u \in H_k^{++}$.

This method is usually called the *beneath-beyond algorithm* for computing the convex hull.

We will illustrate this procedure in the computation of the convex hull for a finite set of points in $\mathbb{R}^2$. Assume for $m > 2$ that

$$Q = \text{conv}(\{v_1, \ldots, v_m\}) = L_1^- \cap \cdots \cap L_m^- \tag{5.9}$$

where $v_i = (x_i, y_i)$ are extreme points of Q,

$$L_i = \{(x, y)^t \in \mathbb{R}^2 \mid a_i x + b_i y = c_i\}$$

and

$$
\begin{aligned}
L_1 \cap L_2 &= \{v_2\} \\
L_2 \cap L_3 &= \{v_3\} \\
&\vdots \\
L_{m-1} \cap L_m &= \{v_m\} \\
L_m \cap L_1 &= \{v_1\}.
\end{aligned}
\tag{5.10}
$$

A simplifying property in $\mathbb{R}^2$ is that for $v \notin Q$, there are at most two pairs of neighboring lines (L, M) from $L_1, \ldots, L_m$ with

$$L \cap M = \{v_i\}, \quad v \in L^{++} \quad \text{and} \quad v \in M^{--} .$$

If $v = (x_0, y_0)$, $L = \{(x, y) \mid a_1 x + b_1 y = c_1\}$ and $M = \{(x, y) \mid a_2 x + b_2 y = c_2\}$, one may check that the double description method adds the new line given by

$$\{(x, y) \mid (\lambda a_2 - \mu a_1)x + (\lambda b_2 - \mu b_1)y = (\lambda c_2 - \mu c_1)\}, \tag{5.11}$$

where $\lambda = a_1 x + b_1 y - c_1 > 0$ and $\mu = a_2 x + b_2 y - c_2 < 0$.

Example 5.14. Consider as a simple example the convex hull Q of $V = \{(0, 0), (1, 0), (1, 1), (0, 1)\} \subseteq \mathbb{R}^2$. Here we may take

$$
\begin{aligned}
L_1 &= \{(x, y) \mid -y = 0\} \\
L_2 &= \{(x, y) \mid x = 1\} \\
L_3 &= \{(x, y) \mid y = 1\} \\
L_4 &= \{(x, y) \mid -x = 0\}.
\end{aligned}
$$

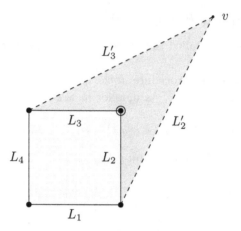

Figure 5.4: Adding the point v to a convex hull in the plane killing the circled extreme point (see Example 5.14).

Adding $v = (2,2)$ to Q, one checks that v belongs to the half planes $L_1^{--}, L_2^{++}, L_3^{++}, L_4^{--}$. Computing $Q' = \mathrm{conv}(Q \cup \{v\})$, using (5.11) L_1 and L_2 combine to give the new line $L_2' = \{(x,y) \mid 2x - y = 2\}$ and L_3 and L_4 combine to give $L_3' = \{(x,y) \mid -x + 2y = 2\}$. Therefore one discards L_2 and L_3 along with $L_2 \cap L_3 = \{(1,1)\}$ and Q' has extreme points $\{(0,0), (1,0), (2,2), (0,1)\}$ with lines L_1, L_2', L_3', L_4 (see Figure 5.4).

The convex hull in Figure 3.6 was computed using this algorithm. Usually the algorithm is too costly without some preprocessing like eliminating points inside the convex hull of four "maximally" chosen points. A famous algorithm for computing the convex hull of n points in the plane was invented by Graham[3] in 1972. Graham notes in the end of [Graham (1972)]: "Computer implementation of this algorithm makes it quite feasible to consider examples with $n = 50000$." perhaps emphasizing that the invention was prompted by practical applications at Bell Labs, where he was employed at the time.

5.4   Linear programming and the simplex algorithm

For historical and perhaps military reasons, the linear optimization problem

$$\max\{c^t x \mid x \in P\}, \tag{5.12}$$

---

[3]Ronald L. Graham (1935–). American mathematician.

where P is a polyhedron, is referred to as a *linear program* (see Figure 5.5) and the activity of solving linear programs as *linear programming*.

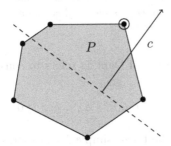

Figure 5.5: A linear program with a planar c and P. The dashed line consists of $v \in \mathbb{R}^2$ with $c^t v = \alpha$ for fixed $\alpha \in \mathbb{R}$. Moving in the direction of c increases the real valued function $f(x) = c^t x$ on $\mathbb{R}^2$. Here f assumes its maximal value at the circled (extreme) point in P.

Mathematically this setup is very simple. It is a real surprise how many interesting problems can be reduced to and solved using linear programs (see §5.4.1). You have already seen a simple example of a linear program in §1.2. There we applied Fourier-Motzkin elimination in finding a solution. In this section we will describe a computational method (the simplex algorithm) observed to be immensely useful for linear programming. The basic idea of the simplex algorithm is touring the vertices of P in a systematic way until an optimal vertex is reached. There is a priori no obvious mathematical reason at all that this approach should be effective.

The simplex algorithm was proposed by George B. Dantzig[4] around 1947. During World War II, Dantzig was in charge of the Air Force Statistical Control's Combat Analysis Branch in the Pentagon. Mechanization of the Air Force's planning process in 1946–47 pointed to the importance of linear programming. Dantzig started approaching Tjalling Koopmans[5] and John von Neumann asking for effective algorithms for linear programming by traversing the vertices of P.

In the beginning this seemed an uphill battle. Dantzig initially considered the algorithm of improving vertices incrementally along edges potentially impossible and kept looking for alternative methods. In his famous monograph [Dantzig (1963)] he writes:

---

[4]George B. Dantzig (1914–2005). American mathematician.
[5]Tjalling Koopmans (1910–1985). Dutch-American mathematical economist.

The obvious idea of moving along edges from one vertex of a convex polyhedron to the next (which underlies the simplex method) was rejected earlier on intuitive grounds as inefficient. In a different geometry it seemed efficient and so, fortunately, it was tested and accepted.

The "different geometry" of Dantzig refers to considering polyhedra on standard form

$$P = \{x \in \mathbb{R}^n \mid Ax = b, x \geq 0\}$$

in (5.12). The usage of the term "simplex" can be traced back to an interpretation of the algorithm in this context. The standard form paved the way for the first computer implementations of the simplex method. For no apparent mathematical reason, these implementations showed the simplex algorithm to be extremely effective. Recently, however, such a mathematical explanation has been given in [Spielman and Teng (2001)].

Here we will introduce the simplex algorithm and the simplex tableau for traversing the vertices of polyhedra not necessarily on standard form. First we give two examples of quite surprising applications of linear programming.

5.4.1   *Two examples of linear programs*

Example 5.15. Let P be the polygon with vertices $(1,1), (2,0), (4,2)$ and $(2,3)$ in $\mathbb{R}^2$. It seems that a unique largest circle can be inscribed in P (see Figure 5.6). What is the center and radius of this circle?

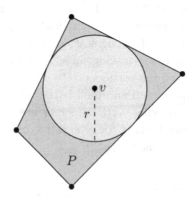

Figure 5.6: A circle centered at v with radius r inside the polygon P.

First notice that $P = \{x \in \mathbb{R}^2 \mid Ax \leq b\}$, where

$$A = \begin{pmatrix} -1 & -1 \\ -2 & 1 \\ 1 & -1 \\ 1 & 2 \end{pmatrix} \quad \text{and} \quad b = \begin{pmatrix} -2 \\ -1 \\ 2 \\ 8 \end{pmatrix}.$$

In general one may prove (see Exercise 5.11) that a ball centered at $v \in \mathbb{R}^n$ with radius r is contained in the affine half plane $H^- = \{x \in \mathbb{R}^n \mid \alpha^t x \leq \beta\}$ if and only if the linear condition

$$\alpha^t v + |\alpha| r \leq \beta$$

holds.

Therefore the general question of how large a ball can be inscribed in a given polyhedron $P = \{x \in \mathbb{R}^n \mid Ax \leq b\}$, can be formulated as the linear program

$$\max\left\{ r \, \middle| \, r \geq 0 \quad \text{and} \quad \begin{matrix} a_1^t v + |a_1| r \leq b_1 \\ \vdots \\ a_m^t v + |a_m| r \leq b_m \end{matrix} \right\} = \max\left\{ r \, \middle| \, \begin{pmatrix} v \\ r \end{pmatrix} \in Q \right\}, \quad (5.13)$$

with

$$Q = \left\{ \begin{pmatrix} v \\ r \end{pmatrix} \, \middle| \, \begin{pmatrix} a_1 & |a_1| \\ \vdots & \vdots \\ a_m & |a_m| \\ 0 & -1 \end{pmatrix} \begin{pmatrix} v \\ r \end{pmatrix} \leq \begin{pmatrix} b_1 \\ \vdots \\ b_m \\ 0 \end{pmatrix} \right\} \subseteq \mathbb{R}^{n+1},$$

where $a_1, \ldots, a_m$ are the rows of A. Notice if v is a vertex of P, then $(v, 0)$ is a vertex of Q. In our concrete example, $(1, 1, 0), (2, 0, 0), (4, 2, 0)$ and $(2, 3, 0)$ are vertices of

$$Q = \left\{ \begin{pmatrix} x \\ y \\ r \end{pmatrix} \, \middle| \, \begin{pmatrix} -1 & -1 & \sqrt{2} \\ -2 & 1 & \sqrt{5} \\ 1 & -1 & \sqrt{2} \\ 1 & 2 & \sqrt{5} \\ 0 & 0 & -1 \end{pmatrix} \begin{pmatrix} x \\ y \\ r \end{pmatrix} \leq \begin{pmatrix} -2 \\ -1 \\ 2 \\ 8 \\ 0 \end{pmatrix} \right\} \subseteq \mathbb{R}^3.$$

The optimal vertex in Q for (5.13) is exactly

$$\left(\frac{10\sqrt{2} + 11\sqrt{5}}{5\sqrt{2} + 4\sqrt{5}}, \frac{3\left(5\sqrt{2} + \sqrt{5}\right)}{5\sqrt{2} + 4\sqrt{5}}, \frac{15}{5\sqrt{2} + 4\sqrt{5}} \right) \in \mathbb{R}^3$$

or approximately $(2.42, 1.74, 0.94)$ corresponding to the circle with radius 0.94 and center $(2.42, 1.74)$. This is the circle depicted in Figure 5.6.

For general polyhedra there may be infinitely many inscribed balls of maximal radius. The solution does not have to be unique as in this case. This illustrates that a linear program may have infinitely many optimal solutions.

Example 5.16. Consider the three points $(1, 2)$, $(2, 1)$ and $(4, 3)$ in $\mathbb{R}^2$ and the problem of finding a best fitting line $y = ax + b$ so that a and b minimize the maximal vertical distance

$$\max\{|2 - a - b|, |1 - 2a - b|, |3 - 4a - b|\} \qquad (5.14)$$

from the line to the points (see Figure 5.7).

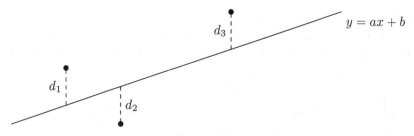

Figure 5.7: The problem of finding a and b, such that the maximal vertical distance $\max\{d_1, d_2, d_3\}$ from three fixed points to the line $y = ax + b$ is minimal, is a linear program. In the classical least squares method one computes a and b minimizing $d_1^2 + d_2^2 + d_3^2$. This can be done with standard linear algebra.

This is an often encountered scenario: the maximal error is subject to minimization. Perhaps this error plays a key role in the stability of a bridge or an airplane. To minimize the maximal error in (5.14) is the same as finding (a, b, d), such that d is minimized subject to

$$|2 - a - b| \leq d$$
$$|1 - 2a - b| \leq d$$
$$|3 - 4a - b| \leq d.$$

Since $|x| \leq d$ if and only if $x \leq d$ and $-x \leq d$ we end up with the linear program $\max\{-d \,|\, (a, b, d) \in P\}$, where

$$P = \left\{ \begin{pmatrix} a \\ b \\ d \end{pmatrix} \in \mathbb{R}^3 \;\middle|\; \begin{pmatrix} -1 & -1 & -1 \\ 1 & 1 & -1 \\ -2 & -1 & -1 \\ 2 & 1 & -1 \\ -4 & -1 & -1 \\ 4 & 1 & -1 \end{pmatrix} \begin{pmatrix} a \\ b \\ d \end{pmatrix} \leq \begin{pmatrix} -2 \\ 2 \\ -1 \\ 1 \\ -3 \\ 3 \end{pmatrix} \right\}.$$

Notice how we (silently) converted the problem of minimizing d into maximizing $-d$. One may compute the optimal line in this case to be $y = \frac{1}{3}x + 1$ (depicted in Figure 5.7) corresponding to $(\frac{1}{3}, 1, \frac{2}{3}) \in P$.

In general one is interested in finding approximate "solutions" to systems of linear equations with small maximum absolute deviation (see Exercise 5.13).

5.4.2 The simplex algorithm in a special case

Before explaining the simplex algorithm for general polyhedra, we will introduce the iterative steps in a special case and illustrate them in a simple example. The simplex algorithm is best understood by carrying out several such computational examples.

Suppose that
$$Q = \{x \in \mathbb{R}^n \mid Ax \le b, x \ge 0\},$$

where A is an $m \times n$ matrix and $b \ge 0$. Then 0 is a vertex of Q by Corollary 4.5. To the linear program $\max\{c^t x \mid x \in Q\}$ with $c \in \mathbb{R}^n$, we associate the $(m + n + 1) \times (n + 1)$ matrix or *tableau*[6]

$$
\begin{array}{c|c}
c^t & 0 \\
\hline
-A & b \\
\hline
I_n & 0
\end{array}
\quad =: \quad
\begin{array}{c|c}
\bar{c} & s \\
\hline
\bar{A} & \bar{b}
\end{array}
\,, \qquad (5.15)
$$

where I_n is the identity $n \times n$ matrix. Here $\bar{A}$ is an $(m + n) \times n$ matrix, $\bar{b} \in \mathbb{R}^{m+n}$ and $\bar{c} \in \mathbb{R}^n$ representing different parts of the tableau in successive iterations of the algorithm. The tableau here is a special case of the general simplex tableau explained in (5.22). In each iteration, the coordinates of the current vertex is available as the last n entries of the last column ($\bar{b}$) in the tableau (5.15). The value of $c^t x$ for the current vertex is in the upper right hand corner (s) of (5.15).

One iteration of the simplex algorithm consists of column operations on the tableau in (5.15) and reflects moving along an edge of Q to a neighboring vertex improving the (cost) function $c^t x$. Here is how this is done.

---

[6]The word *tableau* is French and means picture or table. The plural of tableau is *tableaux*.

(1) If $\overline{c} \leq 0$, the algorithm halts and reports the current vertex as optimal.
(2) If not, $\overline{c}_j > 0$ for some $j = 1, \ldots, n$. Pick such an index j. Let i be the index of a row minimizing

$$-\frac{\overline{b}_k}{\overline{A}_{kj}} \qquad \text{for } \overline{A}_{kj} < 0 \text{ and } k = 1, \ldots, m + n. \qquad (5.16)$$

If no such i exists, the algorithm halts and reports the problem as unsolvable. Otherwise, we refer to $\overline{A}_{ij}$ as the pivot[7] element, i as the pivot row and j as the pivot column.
(3) Normalize the pivot column and use elementary column operations to get zeros in the pivot row (except in the pivot column, where the pivot row contains a 1 from the normalization).
(4) Repeat from (1).

We have deliberately hidden an important point in the above algorithm. In rare cases, the algorithm may cycle indefinitely. One way to prevent this is picking the "smallest" indices j and i in step (2) above. This is made precise in §5.4.5, where the general simplex tableau is introduced.

Example 5.17. Consider the following linear program ([Toft (1988)], Opgave 2.15)

$$\max \quad 5x_1 + 4x_2 + 3x_3$$

subject to

$$
\begin{array}{rcrcrcl}
2x_1 &+& 3x_2 &+& x_3 &\leq& 5 \\
4x_1 &+& x_2 &+& 2x_3 &\leq& 11 \\
3x_1 &+& 4x_2 &+& 2x_3 &\leq& 8 \\
x_1 &&&&&\geq& 0 \\
&& x_2 &&&\geq& 0 \\
&&&& x_3 &\geq& 0.
\end{array}
$$

The initial tableau corresponding to the vertex $(0, 0, 0)$ is:

---

[7]To learn about the etymology of the term "pivot", you are strongly encouraged to take a look at the YouTube video of the illuminating lecture "Tovey explains the column geometry of the simplex method".

$$
\begin{array}{rrr|r}
5 & 4 & 3 & 0 \\
\hline
-2 & -3 & -1 & 5 \\
-4 & -1 & -2 & 11 \\
-3 & -4 & \boxed{-2} & 8 \\
1 & 0 & 0 & 0 \\
0 & 1 & 0 & 0 \\
0 & 0 & 1 & 0
\end{array}
$$

according to (5.15). As a pivot element we have chosen the circled $\overline{A}_{33}$ using (5.16). Normalizing the pivot column in the tableau gives:

$$
\begin{array}{rrr|r}
5 & 4 & -\tfrac{3}{2} & 0 \\
\hline
-2 & -3 & \tfrac{1}{2} & 5 \\
-4 & -1 & 1 & 11 \\
-3 & -4 & 1 & 8 \\
1 & 0 & 0 & 0 \\
0 & 1 & 0 & 0 \\
0 & 0 & -\tfrac{1}{2} & 0
\end{array}
$$

Then elementary column operations are applied to get zeros in the entries of the pivot row outside the pivot position above. This gives the tableau corresponding to the new vertex $(0,0,4)$:

$$
\begin{array}{rrr|r}
\tfrac{1}{2} & -2 & -\tfrac{3}{2} & 12 \\
\hline
-\tfrac{1}{2} & -1 & \tfrac{1}{2} & 1 \\
-1 & 3 & 1 & 3 \\
0 & 0 & 1 & 0 \\
1 & 0 & 0 & 0 \\
0 & 1 & 0 & 0 \\
-\tfrac{3}{2} & -2 & -\tfrac{1}{2} & 4
\end{array}
$$

where the next pivot element $\overline{A}_{11}$ is circled. The exact same procedure gives the tableau:

−1	−3	−1	13
1	0	0	0
2	5	0	1
0	0	1	0
−2	−2	1	2
0	1	0	0
3	1	−2	1

$$(5.17)$$

which is seen to be optimal, since $\bar{c}$ in the upper row is ≤ 0. The optimal vertex $(2, 0, 1)$ is available as the last three coordinates in the $\bar{b}$ part of the tableau in (5.17). The maximum value can be read off from the upper right hand corner in the tableau.

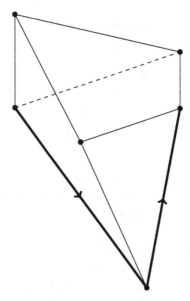

Figure 5.8: Path to optimal solution for the linear program in Example 5.17. Three simplex tableaux are encountered corresponding to the vertex path $(0, 0, 0) \to (0, 0, 4) \to (2, 0, 1)$.

The three-dimensional geometry of the problem and the path to the optimal vertex from $(0, 0, 0)$ is illustrated in Figure 5.8.

The simplex tableau and its operations may seem like magic at this point. They are, however, backed up by simple (but clever!) linear algebra.

We now explain the simplex algorithm for general polyhedra in the context of polyhedral cones and vertices.

5.4.3 *The simplex algorithm for polyhedra in general form*

Suppose that $P = \{x \in \mathbb{R}^n \mid Ax \leq b\}$ in the linear program (5.12) with the assumption that P has a vertex, which is known to us.

If the linear function $c^t x$ is bounded from above on P, there exists a vertex $x_0 \in P$ with

$$c^t x_0 = \max\{c^t x \mid x \in P\} \tag{5.18}$$

by Corollary 4.9. In principle we can solve (5.12) by browsing the vertices of P looking for a vertex x_0, such that $c^t x_0 \geq c^t z$ for every vertex $z \in P$. This potentially involves computing all the vertices of P, which is a daunting task. Fortunately it is possible to decide (locally) if a given vertex is an optimal solution without comparing it to the rest of the vertices.

To approach a local criterion for optimality we introduce the *vertex cone*

$$C_z := \{d \in \mathbb{R}^n \mid A_z d \leq 0\}$$

for a point $z \in P$. We assume here that the reader is familiar with the notation introduced in §4.1.

The extreme rays of the vertex cone correspond to edges in P extending from z. Two vertices $x_0, x_1 \in P$ are called *neighbors* if $\mathrm{conv}(\{x_0, x_1\})$ is an edge of P. In this case, $x_1 - x_0$ is an extreme ray of C_{x_0} (see Exercise 5.14).

The basic geometric idea of the simplex algorithm is moving successively from a vertex $x_0 \in P$ to a neighboring vertex $x_1 \in P$ along an edge, if x_1 improves x_0 i.e., $c^t x_1 > c^t x_0$. If there is no neighboring vertex improving x_0, then x_0 is an optimal solution to (5.12) i.e., x_0 satisfies (5.18). This claim is implied by the following result illustrated in Example 5.19.

Lemma 5.18. *The inclusion*

$$P \subseteq z + C_z$$

holds for every $z \in P$. A point $x_0 \in P$ is optimal for the linear program $\max\{c^t x \mid x \in P\}$ *if and only if*

$$c^t d \leq 0$$

for every $d \in C_{x_0}$.

Proof.   Let $z \in P$. For $x \in P$ we have $A_z x \leq A_z z$ by definition of A_z. This shows that $x - z \in C_z$ and the inclusion $P \subseteq z + C_z$ follows from the identity $x = z + (x - z)$. Therefore $x_0 \in P$ is optimal if $c^t d \leq 0$ for every $d \in C_{x_0}$. Suppose on the other hand that x_0 is optimal and $d \in C_{x_0}$. We know that $x_0 + \epsilon d \in P$ for $\epsilon > 0$ small by definition of C_{x_0}. If $c^t d > 0$ then $c^t(x_0 + \epsilon d) > c^t x_0$ contradicting the optimality of x_0. $\qquad\square$

This rather simple lemma suggests the computations in the simplex algorithm: if x_0 is a vertex of P, then the polyhedral cone C_{x_0} is pointed and therefore generated by its finitely many extreme rays by Theorem 5.3(3). It follows that x_0 fails to be optimal if and only if

$$c^t d > 0 \qquad\qquad (5.19)$$

for some extreme ray $d \in C_{x_0}$. In that case we can move along d to the improving vertex $x_1 = x_0 + \lambda^* d \in \text{ext}(P)$, where

$$\lambda^* = \sup\{\lambda \geq 0 \mid x_0 + \lambda d \in P\}. \qquad\qquad (5.20)$$

Notice that $\lambda^* < \infty$, since $c^t x$ is bounded from above on P. Also $\lambda^* > 0$, since $d \in C_{x_0}$.

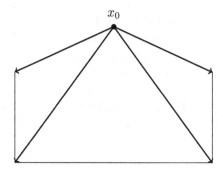

Figure 5.9: The extreme rays of the vertex cone C_{x_0} for $x_0 \in P$ in Example 5.19 sketched as $x_0 + C_{x_0}$.

Example 5.19. Let

$$P = \left\{ \begin{pmatrix} x \\ y \\ z \end{pmatrix} \in \mathbb{R}^3 \,\middle|\, A\begin{pmatrix} x \\ y \\ z \end{pmatrix} \leq b \right\} \subseteq \mathbb{R}^3$$

with

$$A = \begin{pmatrix} 1 & 0 & 1 \\ -1 & 0 & 1 \\ 0 & -1 & 1 \\ 0 & 1 & 1 \\ 0 & 0 & -1 \end{pmatrix} \quad \text{and} \quad b = \begin{pmatrix} 1 \\ 1 \\ 1 \\ 1 \\ 0 \end{pmatrix}.$$

The point $x_0 = (0,0,1)$ is a vertex of P and

$$A_{x_0} = \begin{pmatrix} 1 & 0 & 1 \\ -1 & 0 & 1 \\ 0 & -1 & 1 \\ 0 & 1 & 1 \end{pmatrix}$$

with (see Figure 5.9)

$$C_{x_0} = \{ d \in \mathbb{R}^3 \mid A_{x_0} d \le 0 \} = \text{cone} \left\{ \begin{pmatrix} 1 \\ 1 \\ -1 \end{pmatrix}, \begin{pmatrix} 1 \\ -1 \\ -1 \end{pmatrix}, \begin{pmatrix} -1 \\ -1 \\ -1 \end{pmatrix}, \begin{pmatrix} -1 \\ 1 \\ -1 \end{pmatrix} \right\}.$$

For the linear program $\max\{c^t x \mid x \in P\}$ with $c = (1,1,0)^t$, x_0 fails to be optimal by Lemma 5.18, since $c^t d = 2 > 0$ for

$$d = \begin{pmatrix} 1 \\ 1 \\ -1 \end{pmatrix} \in C_{x_0}.$$

Moving in the direction of d with $\lambda^* = 1$ in (5.20), we get the point $x_0 + d = (1,1,0)$ improving the value of $c^t x$.

5.4.4 *The simplicial hack*

It is much too costly to invoke the double description algorithm (see §5.3) for computing the extreme rays of the vertex cone C_{x_0} in using (5.19) to check that a point $x_0 \in P$ is optimal.

If C_{x_0} happens to be simplicial, the computation of its extreme rays amounts to inversion of a square matrix by Proposition 5.6. For an arbitrary polyhedron P and a vertex $x_0 \in P$, C_{x_0} does not have to be simplicial as illustrated by the cone C_{x_0} in Example 5.19.

The trick of the simplex algorithm is basically relaxing the test in (5.19) to simplicial cones containing the smaller and more complicated vertex cone C_{x_0}. A *simplicial subset* $J \subseteq I(x_0)$ is a subset with $|J| = n$ and A_J invertible. For such a subset we let

$$C_J := \{ d \in \mathbb{R}^n \mid A_J d \le 0 \}.$$

This is a simplicial cone containing C_{x_0}. Notice that $x_0 = A_J^{-1} b_J$ and that the extreme rays in C_J are the columns in the matrix $-A_J^{-1}$ by Proposition 5.6.

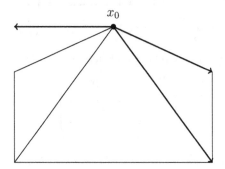

Figure 5.10: $x_0 + C_J$ for $J = \{1, 3, 4\}$ in Example 5.19. The simplicial cone C_J contains C_{x_0} strictly (see Figure 5.9). Notice also that only two of the three extreme rays in C_J correspond to edges in the polyhedron P. Moving along the upper left third one takes us out of P.

Theorem 5.20. *Let x_0 be a vertex of $P = \{x \in \mathbb{R}^n \mid Ax \leq b\}$. Then x_0 is optimal for the linear program*

$$\max\{c^t x \mid x \in P\}$$

if and only if there exists a simplicial subset $J \subseteq I(x_0)$, such that $c^t d \leq 0$ for every $d \in C_J$.

Proof. If there exists a simplicial subset J with these properties, Lemma 5.18 shows that x_0 is optimal. If x_0 is optimal, then $c^t d \leq 0$ for every $d \in C_{x_0}$ by Lemma 5.18. By Lemma 4.21 it follows that

$$c \in \text{cone}(\{a_1, \ldots, a_m\}),$$

where $a_1, \ldots, a_m$ are the rows of A_{x_0}. Now Theorem 3.14 shows that there exists linearly independent rows $a_{i_1}, \ldots, a_{i_n} \in \{a_1, \ldots, a_m\}$ such that

$$c = \lambda_{i_1} a_{i_1} + \cdots + \lambda_{i_n} a_{i_n} \tag{5.21}$$

with $\lambda_{i_j} \geq 0$. With the simplicial subset $J = \{i_1, \ldots, i_n\}$ we have $c^t d \leq 0$ for every $d \in C_J$ by (5.21), since

$$a_{i_j}^t (-A_J^{-1}) = -e_j \leq 0,$$

where e_j is the j-th canonical basis vector in $\mathbb{R}^n$. $\qquad \square$

In the simplex algorithm, a vertex is represented by a simplicial subset J. Moving to a neighboring vertex is the simple operation of exchanging $j \in J$ with $i \notin J$ i.e.,

$$J := (J \setminus \{j\}) \cup \{i\}$$

to get a new simplicial subset under suitable conditions on j and i. Perhaps the biggest surprise about the simplex algorithm is how neatly these operations can be arranged via elementary column operations on a suitable matrix.

5.4.5 The computational miracle of the simplex tableau

Input for the simplex algorithm consists of an $m \times n$ matrix A of rank n with row vectors $a_1, \ldots, a_m$, $b \in \mathbb{R}^m$ and $c \in \mathbb{R}^n$ along with a simplicial subset J, such that

$$x_0 = A_J^{-1} b_J$$

is a vertex of $P = \{x \in \mathbb{R}^n \mid Ax \leq b\}$ in the linear program $\max\{c^t x \mid x \in P\}$.

The *simplex tableau* for the data A, b, c and J is the $(m+1) \times (n+1)$ matrix:

$$
\begin{array}{c|c}
-c^t A_J^{-1} & c^t A_J^{-1} b_J \\
\hline
A A_J^{-1} & b - A A_J^{-1} b_J
\end{array}
\quad =: \quad
\begin{array}{c|c}
\bar{c} & s \\
\hline
\bar{A} & \bar{b}
\end{array}
\tag{5.22}
$$

As indicated above we will use the notation $\bar{c} = -c^t A_J^{-1}$, $\bar{A} = A A_J^{-1}$, $\bar{b} = b - A A_J^{-1} b_J$ and finally $s = c^t A_J^{-1} b_J$ expressing the value of the (cost) function $c^t x$ at the vertex. We will enumerate the columns in A_J^{-1} as $\{d_j \mid j \in J\}$ i.e., d_j is the unique vector $c \in \mathbb{R}^n$ with $c^t a_j = 1$ and $c^t a_k = 0$ for $k \in J \setminus \{j\}$. Notice that $\bar{b} \geq 0$. The current vertex $x_0 = A_J^{-1} b_J$ is optimal if $\bar{c} \leq 0$ in (5.22). This is a consequence of Theorem 5.20 and Proposition 5.6.

Example 5.21. For the simplicial subset $J = \{1, 4, 5\}$ in the linear program in Example 5.16 the simplex tableau is:

$-\frac{1}{3}$	$-\frac{1}{2}$	$-\frac{1}{6}$	$-\frac{2}{3}$
1	0	0	0
$-\frac{1}{3}$	1	$\frac{1}{3}$	$\frac{4}{3}$
$\frac{2}{3}$	0	$\frac{1}{3}$	$\frac{4}{3}$
0	1	0	0
0	0	1	0
$\frac{2}{3}$	1	$-\frac{2}{3}$	$\frac{4}{3}$

Corresponding to the vertex $x_0 = A_J^{-1}b_J = \left(\frac{1}{3}, 1, \frac{2}{3}\right)$. The non-positive entries in $\bar{c}$ show that x_0 is an optimal solution.

If $\bar{c} \not\leq 0$, the current vertex may not be optimal. In that case a move to another simplicial subset is considered. Here are the steps of the simplex algorithm as operations on the tableau in (5.22). The formal details may seem a bit involved. The simplex algorithm is, however, surprisingly easy to carry out for concrete examples (see Example 5.24).

The simplex algorithm

(1) If $\bar{c} \leq 0$, the algorithm halts and reports the current vertex as optimal.
(2) If not, $\bar{c}_j := -c^t d_j > 0$ for some $j \in J$. Pick a *smallest* such j.
(3) Let i be the *smallest* index of a row minimizing

$$-\frac{\bar{b}_k}{\overline{A}_{kj}} \qquad \text{for } \overline{A}_{kj} < 0 \text{ and } k = 1, \dots, m. \qquad (5.23)$$

If no such i exists, the algorithm halts and reports the problem as unsolvable. Otherwise, we refer to $\overline{A}_{ij}$ as the pivot element, i as the pivot row and j as the pivot column in the full simplex tableau (including $\bar{c}, s$ and $\bar{b}$).

(4) $J := (J \setminus \{j\}) \cup \{i\}$.
(5) Normalize the pivot column and use elementary column operations to get zeros in the pivot row using the normalized pivot column.
(6) Repeat from (1).

Explaining the steps

Let d denote $-d_j$ in step ((2)). Here $c^t d > 0$ and we wish to compute

$$\lambda^* = \sup\{\lambda \geq 0 \,|\, x_0 + \lambda d \in P\} = \sup\{\lambda \geq 0 \,|\, A(x_0 + \lambda d) \leq b\}$$
$$= \sup\{\lambda \geq 0 \,|\, \lambda(Ad) \leq \bar{b}\}.$$

In terms of the tableau,

$$\lambda^* = \min\left\{-\frac{\bar{b}_k}{\overline{A}_{kj}} \,\middle|\, \overline{A}_{kj} < 0 \quad k = 1, \ldots, m\right\}. \tag{5.24}$$

If there is no minimizing row for (5.24) i.e., if $\overline{A}_{kj} \geq 0$ for every $k = 1, \ldots, m$, then $x_0 + \lambda d \in P$ for every $\lambda \geq 0$. Since $c^t d > 0$ this implies that $c^t x$ is unbounded on P and the optimization problem is unsolvable. This explains step (3).

If there is a minimizing row for (5.24), the row with the smallest index i is picked and the new simplicial subset

$$J := (J \setminus \{j\}) \cup \{i\}$$

is defined in (4). The vertex $x_0 + \lambda^* d$ is defined by this new simplicial subset. It is important to notice that we may have $\lambda^* = 0$ i.e., it may happen that the simplex algorithm does not move to a neighboring vertex!

The normalization and column operations in (5) of the simplex algorithm need further explanation. Consider the following example illustrating the origin of the necessary updates to the simplex tableau, when changing J into $(J \setminus \{j\}) \cup \{i\}$.

Example 5.22. The two invertible matrices

$$A = \begin{pmatrix} 3 & -1 & -1 \\ -1 & 3 & -1 \\ -1 & -1 & 3 \end{pmatrix} \quad \text{and} \quad B = \begin{pmatrix} -2 & 2 & 6 \\ -1 & 3 & -1 \\ -1 & -1 & 3 \end{pmatrix}$$

differ only in the first row. Suppose that we have computed A^{-1}. Is there a simple way of updating A^{-1} in computing B^{-1} without beginning from scratch? Here the trick is in writing

$$(-2, 2, 6) = \lambda_1(3, -1, -1) + \lambda_2(-1, 3, -1) + \lambda_3(-1, -1, 3)$$

and observing that

$$B = \begin{pmatrix} \lambda_1 & \lambda_2 & \lambda_3 \\ 0 & 1 & 0 \\ 0 & 0 & 1 \end{pmatrix} A.$$

If $\lambda_1 \neq 0$ (in this example $\lambda_1 = 1$),

$$B^{-1} = A^{-1} \begin{pmatrix} \lambda_1 & \lambda_2 & \lambda_3 \\ 0 & 1 & 0 \\ 0 & 0 & 1 \end{pmatrix}^{-1} = A^{-1} \begin{pmatrix} \lambda_1^{-1} & -\lambda_2\lambda_1^{-1} & -\lambda_3\lambda_1^{-1} \\ 0 & 1 & 0 \\ 0 & 0 & 1 \end{pmatrix}.$$

This corresponds to normalizing the first column in A^{-1} by λ_1 and then using elementary column operations multiplying $-\lambda_2$ and $-\lambda_3$ to the first row and adding to the second and third columns respectively.

It is not too difficult to generalize this example into a proof of the following result (see Exercise 5.16).

Proposition 5.23. *Let A be an invertible matrix with rows $a_1, \ldots, a_n \in \mathbb{R}^n$. Suppose that $v = \lambda_1 a_1 + \cdots + \lambda_n a_n$ with $\lambda_j \neq 0$ and let $\tilde{A}$ denote A with the j-th row vector replaced by v. Then $\tilde{A}^{-1}$ can be computed from A^{-1} by dividing the j-th column with λ_j and then using elementary column operations multiplying $-\lambda_1, \ldots, -\lambda_{j-1}, -\lambda_{j+1}, \ldots, -\lambda_n$ to the j-th column and adding to the respective columns $1, \ldots, j-1, j+1, \ldots, n$.*

Suppose that $J = \{i_1, \ldots, i_n\}$ is a simplicial subset. Taking into account that the k-th row in AA_J^{-1} is the coordinates in writing a_k in the basis $a_{i_1}, \ldots, a_{i_n}$, Proposition 5.23 explains the procedure for updating $\overline{A}$ in step (5) of the simplex algorithm. Proposition 5.23 also explains that $\overline{c}$ updates accordingly. The update of $\overline{b}$ by the elementary operation with the pivot column is explained by the identity

$$b - A(A_J^{-1}b_J + \lambda^* d) = \overline{b} - \lambda^* Ad$$

in the notation of (5.24). The update of s in the upper right hand corner is similarly explained by the identity $c^t(x_0 + \lambda^* d) = c^t x_0 - \lambda^*(-c^t d)$.

Notice that we explicitly chose the smallest indices j and i in steps (2) and (3) of the simplex algorithm. There could be several indices j and i to choose from. A rule for choosing i and j is called a *pivot rule*. There are pivot rules that make the simplex algorithm cycle for degenerate input.

The easy pivot rule of picking the smallest indices j and i is called *Bland's rule* after the American mathematician Robert Bland, who discovered it around 1977. With this pivot rule the simplex algorithm is guaranteed to halt (compare this with the algorithm in §3.6).

It is fun to implement the simplex algorithm on a computer. The author programmed it in the language C during an enjoyable Saturday morning.

Most of the effort went into formatting the output. Unlike most modern programming languages, C contains only the most basic instructions.

Example 5.24. Consider the polyhedron

$$P = \left\{ \begin{pmatrix} x \\ y \\ z \end{pmatrix} \in \mathbb{R}^3 \;\middle|\; A\begin{pmatrix} x \\ y \\ z \end{pmatrix} \le b \right\} \subseteq \mathbb{R}^3$$

with

$$A = \begin{pmatrix} 1 & 0 & 1 \\ -1 & 0 & 1 \\ 0 & -1 & 1 \\ 0 & 1 & 1 \\ 0 & 0 & -1 \end{pmatrix} \quad \text{and} \quad b = \begin{pmatrix} 1 \\ 1 \\ 1 \\ 1 \\ 0 \end{pmatrix}$$

and the linear program

$$\max\left\{ -x - y \;\middle|\; \begin{pmatrix} x \\ y \\ z \end{pmatrix} \in P \right\}.$$

The vertex $x_0 = (0,\, 0,\, 1)^t \in P$ has $I(x_0) = \{1, 2, 3, 4\}$.

The simplex tableau corresponding to the simplicial subset $J = \{1, 3, 4\} \subseteq I(x_0)$ is

1	−1	0	0
1	0	0	0
⊝−1	1	1	0
0	1	0	0
0	0	1	0
0	$-\frac{1}{2}$	$-\frac{1}{2}$	1

Here the circled pivot element does not move us away from x_0. The geometric explanation is that we are attempting to follow an improving direction taking us out of P (see Figure 5.11). The pivot step still applies and takes us to another tableau (at the same point!) corresponding to the

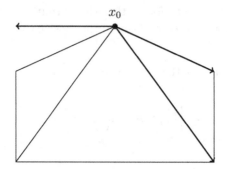

x_0

Figure 5.11: $x_0 + C_J$ for $J = \{1,3,4\}$. The top arrow points out of P, but serves as a stepping stone to the better simplicial cone shown in Figure 5.12.

simplicial subset $J = \{2,3,4\}$:

-1	0	1	0
-1	1	1	0
1	0	0	0
0	1	0	0
0	0	1	0
0	$-\frac{1}{2}$	$\left(-\frac{1}{2}\right)$	1

Here, however, the pivot element leads to the simplicial subset $J = \{2,3,5\}$ and the optimal tableau

-1	-1	-2	2
-1	0	-2	2
1	0	0	0
0	1	0	0
0	-1	-2	2
0	0	1	0

corresponding to the vertex $x = A_J^{-1}b_J = (-1,-1,0)$ (see Figure 5.12).

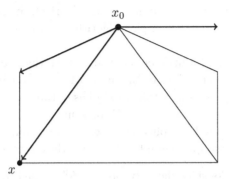

Figure 5.12: $x_0 + C_J$ for $J = \{2, 3, 4\}$. Here C_J contains a direction taking us to the optimal vertex x.

5.4.6 Computing a vertex in a polyhedron

Part of the input for the simplex algorithm is a vertex in $P = \{x \in \mathbb{R}^n \mid Ax \leq b\}$. Often such a vertex comes naturally with the problem at hand, but sometimes it is necessary to make an effort in computing one. Given only A and b in the definition of P, it is not obvious even how to find an element in P or decide if $P = \emptyset$ for that matter.

Here we will assume that A is an $m \times n$ matrix of full rank. In this case P does not contain a line, and we can use the method outlined in §4.2 for computing a vertex given any point in P. Here is a method for computing a point in P using the simplex algorithm with some extra variables.

For the m constraints in $Ax \leq b$ we introduce variables $\epsilon = (\epsilon_1, \ldots, \epsilon_m) \geq 0$ making up for the potential surplus between b and Ax i.e., $Ax - \epsilon \leq b$. The trick is now to consider the linear program

$$\min\{\epsilon_1 + \cdots + \epsilon_m \mid (x, \epsilon) \in Q\}, \tag{5.25}$$

where

$$Q = \{(x, \epsilon) \mid Ax - \epsilon \leq b, \ \epsilon \geq 0\} \subseteq \mathbb{R}^{n+m}. \tag{5.26}$$

The polyhedron Q cannot contain a line, since A is assumed to have full rank. Therefore Q contains a vertex by Theorem 4.7. Let $z_\epsilon := (0, \epsilon) \in \mathbb{R}^{n+m}$ for $\epsilon \in \mathbb{R}^m$. Then $z_\epsilon \in Q$ if

$$\epsilon_i = \begin{cases} 0 & \text{if } b_i \geq 0 \\ -b_i & \text{if } b_i < 0 \end{cases} \tag{5.27}$$

for $i = 1, \ldots, m$. Using z_ϵ as input for the algorithm in §4.2 we may compute a vertex in Q and the simplex algorithm can therefore be applied in solving (5.25).

If you study (5.27) a little closer, you will see that we really only need extra variables corresponding to the indices $1 \le i \le m$ with $b_i < 0$.

The crucial insight is now that (5.25) has optimal value 0 with optimal solution $(x_0, 0)$ if and only if $x_0 \in P$ for some $x_0 \in \mathbb{R}^n$. In this case the algorithm in §4.2 may be applied again to x_0 in computing a vertex of P.

Here is a rather simple example illustrating the essentials of this method.

Example 5.25. Consider the polygon $P \subseteq \mathbb{R}^2$ defined by the inequalities

$$
\begin{aligned}
-x - y &\le -1 \\
x &\le 1 \\
y &\le 1.
\end{aligned}
$$

Corresponding to the first inequality we adjoin the variable $\epsilon \ge 0$. Here the polyhedron $Q \subseteq \mathbb{R}^3$ in (5.26) is given by the inequalities

$$
\begin{aligned}
-x - y - \epsilon &\le -1 \\
x &\le 1 \\
y &\le 1 \\
-\epsilon &\le 0
\end{aligned}
$$

with $z_\epsilon = (0, 0, 1) \in Q$ as in (5.27). Notice that z_ϵ is not a vertex of Q, so the first step is using the algorithm in §4.2 for computing a vertex from z_ϵ. Use the vector $u = (1, -1, 0)$. Here $\lambda = 1$ is the maximal $\lambda > 0$ with $z_\epsilon + \lambda u \in Q$. Proceed with $z_1 := z_\epsilon + u = (1, -1, 1)$ and use $v = (0, 1, -1)$. Here $\lambda = 1$ is the maximal $\lambda > 0$ with $z_1 + \lambda v \in Q$ and $z_2 := v + (0, 1, -1) = (1, 0, 0)$ is finally a vertex of Q. This vertex is optimal for the linear program

$$
\min\{\epsilon \,|\, (x, y, \epsilon) \in Q\}
$$

coming from (5.25). Therefore $(1, 0) \in P$. Since $(1, 0)$ is a vertex of P there is no need for invoking the algorithm in §4.2 once more.

5.5   Exercises

Exercise 5.1. Let A be an $n \times d$ matrix of rank $< d$ and put

$$
C = \{x \in \mathbb{R}^d \,|\, Ax \le 0\}.
$$

(i) Prove that there exists a non-zero vector $z \in \mathbb{R}^d$ with $Az = 0$, $z \in C$ and that such a z is not an extreme ray of C.

(ii) If $z \in \mathbb{R}^d$ is non-zero with $Az = 0$ use the identity

$$r = \tfrac{1}{2}(r + z) + \tfrac{1}{2}(r - z)$$

to prove that C cannot have any extreme rays.

Exercise 5.2. Prove Proposition 5.6 using Theorem 5.3 (1) and the identity

$$AA^{-1} = I_d,$$

where I_d is the $d \times d$ identity matrix.

Exercise 5.3. Use the double description method to verify (5.5).

Exercise 5.4. Use the double description method to verify (5.7).

Exercise 5.5. Check if the inequalities

$$
\begin{aligned}
2x - 3y + z &\leq -2 \\
x + 3y + z &\leq -3 \\
-2x - 3y + z &\leq -2 \\
-x - 3y - 3z &\leq 1 \\
-2x - y + 3z &\leq 3
\end{aligned}
$$

have a solution $x, y, z \in \mathbb{R}$.

Exercise 5.6. Let P be the set of points $(x, y, z) \in \mathbb{R}^3$ satisfying

$$
\begin{aligned}
-x - 2y + 3z &\leq -2 \\
-3x - y + 2z &\leq -1 \\
x &\geq 0 \\
y &\geq 0 \\
z &\geq 0.
\end{aligned}
$$

(i) Compute the extreme points of P.

(ii) Compute the extreme rays in the recession cone of P.

(iii) Write $P = C + Q$ as in (5.1).

Exercise 5.7. Let P denote the set of $(x_1, x_2, x_3, x_4) \in \mathbb{R}^4$ satisfying

$$
\begin{aligned}
x_1 + x_2 + x_3 + x_4 &\geq 1 \\
-3x_1 + x_2 + x_3 + x_4 &\leq 1 \\
x_1 - 3x_2 + x_3 + x_4 &\leq 1 \\
x_1 + x_2 - 3x_3 + x_4 &\leq 1 \\
x_1 + x_2 + x_3 - 3x_4 &\leq 1.
\end{aligned}
$$

Express P as in (5.1). Is P bounded? Does P contain a 4-simplex?

Exercise 5.8. Write down a minimal set of inequalities in x, y and z describing if

$$\begin{pmatrix} x \\ y \\ z \end{pmatrix}$$

is in the convex hull of

$$\begin{pmatrix} 1 \\ 1 \\ 0 \end{pmatrix}, \quad \begin{pmatrix} 1 \\ -1 \\ 0 \end{pmatrix}, \quad \begin{pmatrix} -1 \\ -1 \\ 0 \end{pmatrix}, \quad \begin{pmatrix} -1 \\ 1 \\ 0 \end{pmatrix}, \quad \begin{pmatrix} 0 \\ 0 \\ 1 \end{pmatrix} \quad \text{and} \quad \begin{pmatrix} 0 \\ 0 \\ -1 \end{pmatrix}$$

in $\mathbb{R}^3$. A minimal inequality corresponds to a facet. Show that the convex hull of five points not all in the same affine hyperplane of $\mathbb{R}^3$ has at most six facets.

Exercise 5.9. Compute $\mathrm{rec}(P)$, where P is the set of solutions to

$$\begin{array}{rrrrr}
-x & - & y & + & z & \leq & 1 \\
-x & + & 2y & - & z & \leq & 2 \\
x & - & 2y & + & z & \leq & 1 \\
5x & + & 5y & - & 7z & \leq & 5
\end{array}$$

in $\mathbb{R}^3$.

Exercise 5.10.

(i) Let P denote the set of $(x, y, z) \in \mathbb{R}^3$ satisfying

$$\begin{array}{rrrrr}
x & + & y & + & z & \leq & 1 \\
x & + & y & - & z & \leq & 1 \\
x & - & y & - & z & \leq & 1 \\
-x & + & y & - & z & \leq & 1 \\
x & - & y & + & z & \leq & 1 \\
-x & + & y & + & z & \leq & 1 \\
-x & - & y & + & z & \leq & 1 \\
-x & - & y & - & z & \leq & 1.
\end{array}$$

Show that P is the convex hull of 6 points.

(ii) Show that P has 12 one-dimensional faces (edges).

Exercise 5.11. The ball centered at $v \in \mathbb{R}^n$ with radius $r \geq 0$ is the subset

$$B(v, r) := \left\{ x \in \mathbb{R}^n \,\middle|\, |x - v|^2 \leq r^2 \right\}.$$

Let $H^- = \{ x \in \mathbb{R}^n \mid \alpha^t x \leq \beta \}$ be a given affine half space.

(i) Prove that

$$B(v, r) = \{v + \lambda u \,|\, 0 \le \lambda \le r, |u| = 1\}.$$

(ii) Prove that $B(v, r) \subseteq H^-$ if and only if

$$\alpha^t v + |\alpha| r \le \beta.$$

Exercise 5.12. It also seems that there exists a circle of minimal radius containing the polygon in Example 5.15. Compute the center and radius of this circle.

Exercise 5.13. It may happen that a system

$$a_{11} x_1 + \cdots + a_{1n} x_n = b_1$$
$$\vdots$$
$$a_{m1} x_1 + \cdots + a_{mn} x_n = b_m$$

of linear equations does not have a solution. In practice one is interested in an approximate "solution", such that the maximal absolute error

$$\max \{|b_1 - (a_{11} x_1 + \cdots + a_{1n} x_n)|, \ldots, |b_m - (a_{m1} x_1 + \cdots + a_{mn} x_n)|\}$$

is minimal. Give a suggestion as to how this problem can be solved.

Exercise 5.14. Let $x, y \in P \subseteq \mathbb{R}^n$ be vertices in a polyhedron P. Prove that $y - x$ is an extreme ray in C_x if $\mathrm{conv}\{x, y\}$ is an edge in P. Let $d \in C_x$ be an extreme ray. Prove that $\{x + \lambda d \,|\, \lambda \ge 0\} \cap P$ is an edge of P. Does an extreme ray in C_x necessarily lead to a neighboring vertex?

Exercise 5.15. Explain why the tableau given in (5.15) is a special case of the general simplex tableau in (5.22). Why are the last n entries in the $\bar{b}$-column of (5.15) the coordinates of the current vertex?

Exercise 5.16. Give a proof of Proposition 5.23.

Exercise 5.17. Compute an optimal solution of the linear program

$$\begin{aligned}
\max \quad & x_1 + x_2 + x_3 \\
\text{subject to} \\
-x_1 + x_2 + x_3 &\le 1 \\
x_1 - x_2 + x_3 &\le 2 \\
x_1 + x_2 - x_3 &\le 3 \\
x_1 &\ge 0 \\
x_2 &\ge 0 \\
x_3 &\ge 0
\end{aligned}$$

using the simplex algorithm. Is your optimal solution unique?

Exercise 5.18. Let $P(\lambda)$ denote the linear program

$$\max \quad x_1 + \lambda x_2$$

subject to

$$
\begin{aligned}
\tfrac{1}{2}x_1 + x_2 &\leq 5 \\
x_1 + x_2 &\leq 6 \\
2x_1 - x_2 &\leq 4 \\
x_1 &\geq 0 \\
x_2 &\geq 0
\end{aligned}
$$

for $\lambda \in \mathbb{R}$. Compute an optimal solution for $P(0)$. Find the optimal solutions for $P(\lambda)$ in general for $\lambda \in \mathbb{R}$. The problem in this exercise falls under the heading *parametric linear programming*. This means solving linear programs depending on a parameter λ.

Chapter 6

Closed convex subsets and separating hyperplanes

A planar disc is a convex subset with infinitely many extreme points and therefore not the intersection of finitely many affine half planes (see Figure 6.1).

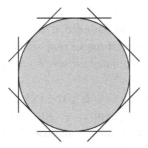

Figure 6.1: A planar disc is not the intersection of finitely many affine half planes.

In this chapter we consider convex subsets that are intersections of possibly infinitely many affine half spaces. It is proved in Theorem 6.5 that these convex subsets coincide with the *closed convex subsets*.

Recall that a subset $F \subseteq \mathbb{R}^d$ is called *closed* (see §A.4) if every convergent sequence $(x_n) \subseteq F$ has its limit inside F. The notion of a subset being closed comes from analysis. Sometimes I present the following example to check that students can work with and not just recite this definition. If it puzzles you, it may be a good idea to refresh notions studying the appendix (Appendix A) on analysis. If you consider it an insult or too easy, then read on.

Example 6.1. The subset $S = \{x \in \mathbb{R} \mid x > 0\} \subseteq \mathbb{R}$ is not a closed subset as (x_n) with $x_n = 1/n$ is a convergent sequence with $x_n \in S$, but the limit of (x_n) is 0, which is not in S.

A first result is that closed convex subsets really are intersections of affine half spaces. They can be considered as solution sets for infinitely many linear inequalities as opposed to the finitely many for polyhedra.

The main result in this chapter is the existence of a supporting hyperplane at a boundary point of a convex subset (Theorem 6.11). As an application of the main result we prove two results both due to Minkowski. The first (Theorem 6.12) states that every convex compact subset is the convex hull of its extreme points. The second (Theorem 6.15) states that disjoint convex subsets can be separated by an affine hyperplane (see §6.3).

6.1   Closed convex subsets

Recall that a sequence (x_n) is *bounded* if there exists $M > 0$, such that $|x_n| \leq M$ for every n. A cornerstone in real analysis is the surprising result (see Theorem A.13) that a *bounded sequence* (x_n) always has a *convergent subsequence*.

For a closed convex subset $C \subseteq \mathbb{R}^d$, the following result says that there always exists a unique closest point $x_0 \in C$ to any given point $y \in \mathbb{R}^d$. Furthermore this point has a very nice geometric description (see Figure 6.2).

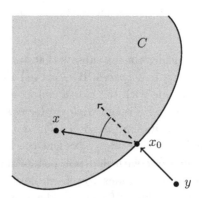

Figure 6.2: The geometric characterization of the closest point x_0 in a closed convex subset C to a point $y \notin C$: the angle between $x_0 - y$ and $x - x_0$ must be acute for every $x \in C$.

Theorem 6.2. *Let $C \subseteq \mathbb{R}^d$ be a closed convex subset and $y \in \mathbb{R}^d$. Then there exists a point $x_0 \in C$ closest to y i.e.,*

$$|x_0 - y| \leq |x - y|$$

for every $x \in C$. This point is uniquely given by the property that

$$(x - x_0)^t (x_0 - y) \geq 0 \tag{6.1}$$

for every $x \in C$.

Proof. Let

$$m = \inf \{ |y - x| \,|\, x \in C \}.$$

By definition (see §A.2.1), there exists for every n, $x_n \in C$ such that $|y - x_n| < m + 1/n$. This shows that (x_n) is a bounded sequence of elements in C. Let (x_{n_i}) be a convergent subsequence with limit x_0. Then $x_0 \in C$, since C is closed. We must have $|x_0 - y| = m$. Therefore $|x_0 - y| \leq |x - y|$ for every $x \in C$.

Suppose now that $(x - x_0)^t (x_0 - y) \geq 0$ for every $x \in C$. Then

$$\begin{aligned} |x - y|^2 &= |x - x_0 + x_0 - y|^2 \\ &= |x - x_0|^2 + |x_0 - y|^2 + 2(x - x_0)^t (x_0 - y). \end{aligned}$$

Therefore $|x - y|^2 - |x_0 - y|^2 \geq 0$ for every $x \in C$ and $x_0 \in C$ is a closest point in C to y. On the other hand, if $x \in C$,

$$\begin{aligned} |(1 - \lambda)x_0 + \lambda x - y|^2 &= |x_0 + \lambda(x - x_0) - y|^2 \\ &= |x_0 - y|^2 + \lambda^2 |x - x_0|^2 + 2\lambda(x - x_0)^t (x_0 - y) \end{aligned}$$

and since x_0 is a closest point in C to y and $(1 - \lambda)x_0 + \lambda x \in C$,

$$\begin{aligned} |(1 - \lambda)x_0 + \lambda x - y|^2 &- |x_0 - y|^2 \\ &= \lambda(\lambda |x - x_0|^2 + 2(x - x_0)^t (x_0 - y)) \geq 0 \end{aligned}$$

for every $0 \leq \lambda \leq 1$. Consequently

$$\lambda |x - x_0|^2 + 2(x - x_0)^t (x_0 - y) \geq 0$$

for every $0 < \lambda \leq 1$. This implies that $(x - x_0)^t (x_0 - y) \geq 0$. The uniqueness of $x_0 \in C$ is left as Exercise 6.3. $\qquad\square$

Remark 6.3. The converse of Theorem 6.2 also holds: if $C \subseteq \mathbb{R}^d$ is a non-empty subset with the property that to each point of $\mathbb{R}^d$ there is a unique closest point in C, then C is closed and convex. This result is due to Bunt[1](1934) and Motzkin (1935).

Corollary 6.4. *Let $C \subseteq \mathbb{R}^d$ be a closed convex subset and suppose that $y \notin C$. Then there exists an affine hyperplane*

$$H = \{x \in \mathbb{R}^d \,|\, \alpha^t x = \beta\},$$

such that $C \subseteq H^+$ and $y \notin H^+$ i.e., $\alpha^t x \geq \beta$ for every $x \in C$ and $\alpha^t y < \beta$.

Proof.   Let $x_0 \in C$ be the point closest to y according to Theorem 6.2. Put $\alpha = x_0 - y$ and $\beta = x_0^t \alpha$. Then the inclusion $C \subseteq H^+$ follows from (6.1). Since $y \neq x_0$ we must have $\alpha^t(x_0 - y) > 0$. Therefore $\alpha^t y < \beta$.   □

With this observation we get one of the key properties of closed convex subsets.

Theorem 6.5. *A closed convex subset $C \subseteq \mathbb{R}^d$ is the intersection of the affine half spaces containing it.*

Proof.   Clearly C is contained in the intersection, since it is contained in each of the half spaces. If $C = \mathbb{R}^d$, C is the trivial (empty) intersection of affine half spaces. Suppose that $C \neq \mathbb{R}^d$. If y is a point in the intersection outside C, there exists a hyperplane H with $C \subseteq H^+$ and $y \notin H^+$ by Corollary 6.4. This contradicts that y is in the intersection of all half spaces containing C.   □

Theorem 6.5 tells an important story about closed convex subsets. An affine half space $H^+ \subseteq \mathbb{R}^d$ is the set of solutions to a linear inequality. Therefore a closed convex subset really is the set of common solutions to a (possibly infinite) set of linear inequalities.

Recall that the polar of a cone $C \subseteq \mathbb{R}^d$ is

$$C^\circ = \{\alpha \in \mathbb{R}^d \,|\, \alpha^t x \leq 0, \text{for every } x \in C\}.$$

For closed convex cones we have the following strengthening of Theorem 6.5.

Lemma 6.6. *Let $C \subseteq \mathbb{R}^d$ be a closed convex cone. Then*

$$C = \bigcap_{\alpha \in C^\circ} \{x \in \mathbb{R}^d \,|\, \alpha^t x \leq 0\}.$$

---

[1]Lucas Nicolaas Hendrik Bunt (1905–1984). Dutch mathematician.

Proof. If $C \subseteq H^-$ with $H = \{x \in \mathbb{R}^d \mid \alpha^t x = \beta\}$, we must have $\beta \geq 0$ as $0 \in C$ and also $\alpha^t x \leq 0$ for every $x \in C$, since a cone is stable under dilation. Now the result follows from Theorem 6.5. $\qquad \square$

As a consequence of Lemma 6.6 we have the following analogue of Farkas's lemma (Lemma 4.14) illustrating why "closed" is the perfect infinite generalization of "finitely generated".

Lemma 6.7. *Suppose that $C \subseteq \mathbb{R}^d$ is a closed convex cone. Then*

(1) *If $x \notin C$, there exists $\alpha \in C^\circ$ with $\alpha^t x > 0$.*
(2) *$(C^\circ)^\circ = C$.*

6.2 Supporting hyperplanes

With some more attention to detail we can actually prove that any (not necessarily closed) convex subset $C \subsetneq \mathbb{R}^d$ is contained in one side of an affine hyperplane "touching" C at its boundary. Recall the definition of the *boundary* ∂S of a subset $S \subseteq \mathbb{R}^d$ (see §A.5). This is the set of points which are at the same time limit points for convergent sequences with elements from S and convergent sequences with elements from the complement of S. To put it briefly,

$$\partial S = \overline{S} \cap \overline{\mathbb{R}^d \setminus S}.$$

Proposition 6.8. *Let $C \subseteq \mathbb{R}^d$ be a convex subset. Then the closure, $\overline{C}$, of C is a convex subset.*

Proof. Consider $x, y \in \overline{C}$. We must prove that $z := (1 - \lambda)x + \lambda y \in \overline{C}$ for $0 \leq \lambda \leq 1$. By definition of the closure $\overline{C}$ there exist convergent sequences (x_n) and (y_n) with $x_n, y_n \in C$ such that $x_n \to x$ and $y_n \to y$. Now form the sequence $((1 - \lambda)x_n + \lambda y_n)$. Since C is convex this is a sequence of vectors in C. The convergence of (x_n) and (y_n) allows us to conclude that

$$(1 - \lambda)x_n + \lambda y_n \to z.$$

Since z is the limit of a convergent sequence with vectors in C, we have shown that $z \in \overline{C}$. $\qquad \square$

It is not too hard to give an example of a proper subset S of $\mathbb{R}^d$ with $\overline{S} = \mathbb{R}^d$. However, if C is a proper convex subset of $\mathbb{R}^d$, then $\overline{C}$ is a proper subset of $\mathbb{R}^d$. This is intuitively quite clear, but it calls for a proof (try to do Exercise 6.9 before reading on).

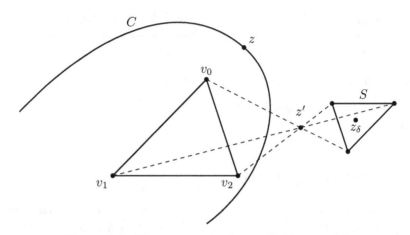

Figure 6.3: The geometry in the proof of Lemma 6.9.

We need the following very intuitive result[2], which has an interesting geometric proof (see Figure 6.3) essentially mirroring a simplex inside a convex subset to one on the outside. It shows that points on the boundary of a proper convex subset can be approximated from outside its closure.

Lemma 6.9. *Let C be a convex subset of $\mathbb{R}^d$. If $z \in \partial C$, then for every $\epsilon > 0$, there exists $y \notin \overline{C}$ with $|y - z| < \epsilon$.*

Proof. If $\dim C < d$, then C is contained in an affine hyperplane H. Since $\overline{C} \subseteq H$, we have $\mathbb{R}^d \setminus H \subseteq \mathbb{R}^d \setminus \overline{C}$. But $\overline{\mathbb{R}^d \setminus H} = \mathbb{R}^d$ and the result follows in this case. If $\dim C = d$ we can find $d + 1$ affinely independent points $v_0, \dots, v_d \in C$. Since $z \in \partial C$, there exists $z' \notin C$ with $|z' - z| < \epsilon/2$. For $\delta > 0$ we let (see Figure 6.3)

$$S = \mathrm{conv}((1 + \delta)z' - \delta v_0, \dots, (1 + \delta)z' - \delta v_d).$$

Since

$$z' = \frac{\delta}{1 + \delta}(\lambda_0 v_0 + \cdots + \lambda_d v_d)$$
$$+ \frac{1}{1 + \delta}\Big(\lambda_0\big((1 + \delta)z' - \delta v_0\big) + \cdots + \lambda_d\big((1 + \delta)z' - \delta v_d\big)\Big)$$

---

[2]For closed convex subsets this result is not needed. See Exercise 6.22.

whenever $\lambda_i \geq 0$ and $\lambda_0 + \cdots + \lambda_d = 1$, it follows that $S \cap C = \emptyset$. By Exercise 6.6,

$$z_\delta = (1+\delta)z' - \frac{\delta}{d+1}(v_0 + \cdots + v_d)$$

is an interior point of S. Therefore $z_\delta \notin \overline{C}$. We may choose $\delta > 0$, such that $|z' - z_\delta| < \epsilon/2$. Therefore $|z - z_\delta| \leq |z - z'| + |z' - z_\delta| < \epsilon$ and $y = z_\delta$ satisfies the requirement. $\qquad\square$

Definition 6.10. A *supporting hyperplane* for a convex subset $C \subseteq \mathbb{R}^d$ at a boundary point $z \in \partial C$ is an affine hyperplane H with $z \in H$ and $C \subseteq H^+$.

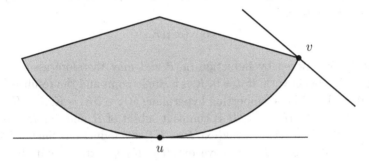

Figure 6.4: A convex subset $C \subseteq \mathbb{R}^2$ along with supporting hyperplanes at two of its boundary points u and v. There is a unique supporting hyperplane at u, but infinitely many at v.

Theorem 6.11. *Let $C \subseteq \mathbb{R}^d$ be a convex subset and $z \in \partial C$. Then there exists a supporting hyperplane for C at z.*

Proof. By Lemma 6.9, there exists a sequence of points (z_n) with $z_n \notin \overline{C}$, such that $z_n \to z$. Proposition 6.8 says that $\overline{C}$ is a convex subset. Therefore Theorem 6.2 implies the existence of $x_n \in \overline{C}$, such that $|z_n - x_n| \leq |z_n - x|$ and $(x_n - z_n)^t(x - x_n) \geq 0$ for every $x \in \overline{C}$. Now let

$$u_n = \frac{x_n - z_n}{|x_n - z_n|}.$$

Then

$$u_n^t(x - x_n) \geq 0 \tag{6.2}$$

for every $x \in \overline{C}$. Since (u_n) is a bounded sequence, it has a convergent subsequence. Let u be the limit of this convergent subsequence. Then (6.2) shows that $H = \{x \in \mathbb{R}^n \mid u^t x = u^t z\}$ is a supporting hyperplane for C at z, since $x_n \to z$ as $n \to \infty$. This follows from the inequality

$$|z - x_n| \leq |z - z_n| + |z_n - x_n| \leq 2|z - z_n|,$$

where we have used that $|z_n - x_n| \leq |z_n - z|$. $\qquad\qquad\square$

The existence of a supporting hyperplane at a boundary point of a convex subset has real content. It is the key ingredient in the proof of the following classical result.

Theorem 6.12 (Minkowski). *Let K be a compact convex subset of $\mathbb{R}^d$. Then* $\mathrm{ext}(K) \neq \emptyset$ *and*

$$K = \mathrm{conv}(\mathrm{ext}(K)).$$

Proof.   We proceed by induction on d and may therefore assume that $\dim K = d$ to begin with. If $d = 0$, K is a single point and the result is true. If $d > 0$, we let H be a supporting hyperplane at $z \in \partial K = K \setminus \mathrm{int}(K)$ (see Exercise 6.20). As $H \cap K \subseteq H$ is compact subset of H and $\dim H = d - 1$ it follows that $H \cap K = \mathrm{conv}(\mathrm{ext}(H \cap K))$ by induction. Since $H \cap K$ is an exposed face of K, we have $\mathrm{ext}(H \cap K) \subseteq \mathrm{ext}(K)$ and therefore $x \in \mathrm{conv}(\mathrm{ext}(K))$ if $x \in \partial K$. Suppose that $x \in \mathrm{int}(K)$ and $v \in \mathbb{R}^d \setminus \{0\}$. Since K is compact both $t^- = \max\{t \mid x - tv \in K\}$ and $t^+ = \max\{t \mid x + tv \in K\}$ exist and $x - t^- v, x + t^+ v \in \partial K$. Since

$$x = \frac{t^+}{t^- + t^+}(x - t^- v) + \frac{t^-}{t^- + t^+}(x + t^+ v),$$

x is a convex combination of two points in ∂K and the proof is complete. $\quad\square$

Here is a generalization of Corollary 4.9 showing that a linear function on a compact convex subset attains its maximum in an extreme point.

Corollary 6.13. *Let $K \subseteq \mathbb{R}^d$ be a compact convex subset and $c \in \mathbb{R}^d$. If $z \in K$ satisfies $c^t z \geq c^t x$ for every $x \in K$, there exists $x_0 \in \mathrm{ext}(K)$ with $c^t x_0 = c^t z$.*

Proof.   The proof is similar to the proof of Corollary 4.9 (see Exercise 6.11). $\qquad\qquad\square$

6.3   Separation by hyperplanes

Two subsets S_1 and S_2 of $\mathbb{R}^d$ are *separated* by an affine hyperplane H if

$$S_1 \subseteq H^- \quad \text{and} \quad S_2 \subseteq H^+. \tag{6.3}$$

The separation (6.3) is called *proper* if $S_1 \cup S_2 \not\subseteq H$, *strict* if $S_1 \cap H = \emptyset$ and $S_2 \cap H = \emptyset$ and *strong* if there exists $\epsilon > 0$, such that the two convex subsets $S_1 + \epsilon B$ and $S_2 + \epsilon B$ are strictly separated. Here B denotes the unit ball $\{x \in \mathbb{R}^d \,|\, |x| \leq 1\}$ in $\mathbb{R}^d$.

Example 6.14. With these new terms, Theorem 6.2 implies that a closed convex subset C can be strongly separated from $\{y\}$ if $y \notin C$.

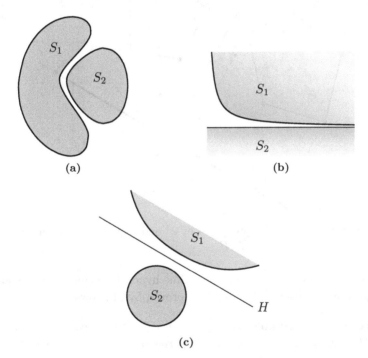

(a)   (b)

(c)

Figure 6.5: Examples of subsets $S_1, S_2 \subseteq \mathbb{R}^2$ such that, (a) S_1 and S_2 cannot be separated (by an affine hyperplane), (b) Here $S_1 = \{(x, y) \,|\, x > 0, \ y \geq 1/x\}$ and $S_2 = \{(x, y) \,|\, x \geq 0, \ y = 0\}$ are disjoint strictly separated convex subsets that cannot be strongly separated, (c) S_1 and S_2 are convex subsets strongly separated by H.

Separation of convex subsets by hyperplanes may seem quite abstract and made up. A prominent application is in collision detection in robotics and computer games. You have some representations of two convex subsets S_1 and S_2 in $\mathbb{R}^3$ or $\mathbb{R}^2$ and wish to detect if $S_1 \cap S_2 = \emptyset$ signifying that S_1 and S_2 have not collided. The content of Theorem 6.15 says that this can be detected by projection onto for example a line if $S_1, S_2 \subseteq \mathbb{R}^2$ (see Figure 6.6). This line is normal to a supporting hyperplane of S_1 or S_2. If S_1 and S_2 are polyhedra this leads to an explicit algorithm traversing the normal vectors of edges of S_1 and S_2.

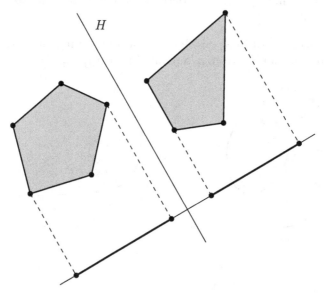

Figure 6.6: Two disjoint polyhedra in $\mathbb{R}^2$ and their projection onto a line normal to a separating axis H.

It seems intuitively clear that an affine hyperplane can always separate disjoint convex subsets. This was first proved by Minkowski.

Theorem 6.15 (Minkowski). *Let $C_1, C_2 \subseteq \mathbb{R}^d$ be disjoint convex subsets. Then there exists a separating hyperplane*

$$H = \{x \in \mathbb{R}^d \,|\, \alpha^t x = \beta\}$$

for C_1 and C_2. If in addition, C_1 is compact and C_2 is closed, then C_1 and C_2 can be strongly separated.

Proof.   The key idea in the proof is the existence of a non-zero $\alpha \in \mathbb{R}^d$, such that $\alpha^t x \geq \alpha^t y$ for every $x \in C_1$ and every $y \in C_2$. The trick lies in considering the convex subset

$$C_1 - C_2 = \{x - y \mid x \in C_1, y \in C_2\} \subseteq \mathbb{R}^d.$$

Here $0 \notin C_1 - C_2$ since $C_1 \cap C_2 = \emptyset$. The closure $\overline{C_1 - C_2}$ is convex by Proposition 6.8. If $0 \notin \overline{C_1 - C_2}$, there exists $\alpha \in \mathbb{R}^d$ and $\epsilon \in \mathbb{R}$

$$\alpha^t(x - y) > \epsilon > 0 \tag{6.4}$$

for every $x \in C_1, y \in C_2$ by Theorem 6.2.

If $0 \in \overline{C_1 - C_2}$ the existence of a non-zero α with $\alpha^t(x - y) \geq 0$ for every $x \in C_1$ and every $y \in C_2$ follows from Theorem 6.11 with $z = 0$. Thus there exists a non-zero $\alpha \in \mathbb{R}^d$ with $\alpha^t x \geq \alpha^t y$ for every $x \in C_1$ and every $y \in C_2$. Putting

$$\beta = \sup\{\alpha^t y \mid y \in C_2\}$$

it follows by the definition of supremum (see §A.2.1) that $\alpha^t x \geq \beta$ for every $x \in C_1$. Therefore

$$H = \{x \in \mathbb{R}^d \mid \alpha^t x = \beta\}$$

is the desired separating hyperplane with $C_1 \subseteq H^+$ and $C_2 \subseteq H^-$.

Assume now that C_1 is compact and C_2 is closed. Then we claim that $0 \notin \overline{C_1 - C_2}$. If $0 \in \overline{C_1 - C_2}$, then there exists a sequence $(x_n - y_n)$ converging to 0 with $x_n \in C_1$ and $y_n \in C_2$. Since C_1 is compact, (x_n) has a convergent subsequence (x_{n_j}) converging to $x \in C_1$. Since $(x_{n_j} - y_{n_j})$ converges to 0, this implies that the sequence (y_{n_j}) is convergent with limit $y \in \mathbb{R}^d$. As C_2 is closed we have $y \in C_2$. Therefore $x - y = 0$ contradicting that $C_1 \cap C_2 = \emptyset$ and (6.4) shows strong separation between C_1 and C_2. $\qquad\square$

6.4 Exercises

Exercise 6.1. Prove that

$$K = \{(x, y) \in \mathbb{R}^2 \mid x > 0, \, xy \geq 1\}$$

is a convex and closed subset of $\mathbb{R}^2$.

Exercise 6.2. Give examples of convex subsets that are

    (i) bounded
    (ii) unbounded
    (iii) closed
    (iv) open
    (v) neither open nor closed

in $\mathbb{R}^2$.

Exercise 6.3. Let $x_0, x_0', y \in \mathbb{R}^n$. Prove that if

$$(x_0' - x_0)^t (x_0 - y) \geq 0$$
$$(x_0 - x_0')^t (x_0' - y) \geq 0,$$

then $x_0 = x_0'$.

Exercise 6.4. Can you prove that

    (i) a polyhedral cone is a closed subset
    (ii) a finitely generated cone is a closed subset

from first principles i.e., only using the definitions?

Exercise 6.5. Let $F_1, F_2 \subseteq \mathbb{R}^2$ be closed subsets. Is

$$F_1 - F_2 = \{x - y \mid x \in F_1, \, y \in F_2\}$$

a closed subset of $\mathbb{R}^2$?

Exercise 6.6. In Exercise 2.11 it is proved that

$$f(x) = (\lambda_0, \lambda_1, \ldots, \lambda_d)$$

is a well defined affine map $f : \mathbb{R}^d \to \mathbb{R}^{d+1}$, where

$$x = \lambda_0 v_0 + \cdots + \lambda_d v_d$$

with $v_0, \ldots, v_d$ affinely independent points in $\mathbb{R}^d$ and $\lambda_0 + \cdots + \lambda_d = 1$. Use this along with the open set

$$U = \{(\lambda_0, \ldots, \lambda_d) \in \mathbb{R}^{d+1} \mid \lambda_0, \ldots, \lambda_d > 0\}$$

to prove that

$$\frac{1}{d+1}(v_0 + v_1 + \cdots + v_d)$$

is an interior point of S in $\mathbb{R}^d$, where S is the $(d+1)$-simplex

$$S = \text{conv}(\{v_0, v_1, \ldots, v_d\}).$$

Exercise 6.7. Prove that $\text{conv}(\{v_1, \ldots, v_m\})$ is a compact subset of $\mathbb{R}^n$ by considering the continuous map $f : \mathbb{R}^m \to \mathbb{R}^n$ given by

$$f(t_1, \ldots, t_m) = t_1 v_1 + \cdots + t_m v_m$$

along with $f(\Delta)$, where $\Delta = \{(t_1, \ldots, t_m)^t \in \mathbb{R}^m \mid t_1 + \cdots + t_m = 1, t_1 \geq 0, \ldots, t_m \geq 0\}$.

Hint: seek inspiration in §A.7.

Exercise 6.8. Let K be a compact subset of $\mathbb{R}^n$. Prove that $\text{conv}(K)$ is compact by using Corollary 3.15. You may find it useful to look at Exercise 6.7.

Exercise 6.9. Let $C \subseteq \mathbb{R}^d$ be a convex subset. Prove that $\overline{C} \neq \mathbb{R}^d$ if $C \neq \mathbb{R}^d$.

Exercise 6.10. Find all the supporting hyperplanes of the triangle with vertices $(0,0), (0,2)$ and $(1,0)$.

Exercise 6.11. Complete the proof of Corollary 6.13.

Exercise 6.12. Give an example of a non-proper separation of convex subsets.

Exercise 6.13. Prove that the two subsets in Figure 6.5(b) are strictly separated closed convex subsets that cannot be separated strongly.

Exercise 6.14. In Theorem 6.15 it is proved that two disjoint convex subsets C_1 and C_2 of $\mathbb{R}^n$ can be separated by an affine hyperplane H i.e., $C_1 \subseteq H^-$ and $C_2 \subseteq H^+$. Can this result be strengthened to $C_1 \cap H = \emptyset$? If not, give a counterexample.

Exercise 6.15. Let

$$B_1 = \{(x,y) \mid x^2 + y^2 \leq 1\}$$
$$B_2 = \{(x,y) \mid (x-2)^2 + y^2 \leq 1\}$$

(i) Show that B_1 and B_2 are closed convex subsets of $\mathbb{R}^2$.
(ii) Find a hyperplane properly separating B_1 and B_2.

(iii) Can you separate B_1 and B_2 strictly?
(iv) Put $B_1' = B_1 \setminus \{(1,0)\}$ and $B_2' = B_2 \setminus \{(1,0)\}$. Show that B_1' and B_2' are convex subsets. Can you separate B_1' from B_2 strictly? What about B_1' and B_2'?

Exercise 6.16. Let $C = \{(x,y) \in \mathbb{R}^2 \,|\, (x-1)^2 + y^2 \leq 1\}$ and $v = (0,2)$. What is the point in C closest to v? Find the equation of a hyperplane separating $\{v\}$ from C strictly.

Exercise 6.17. Let S be the square with vertices $(0,0)$, $(1,0)$, $(0,1)$ and $(1,1)$ and let $P = (2,0)$.

(i) Find the set of hyperplanes through $(1,\frac{1}{2})$, which separate S from P.
(ii) Find the set of hyperplanes through $(1,0)$, which separate S from P.
(iii) Find the set of hyperplanes through $(\frac{3}{2},1)$, which separate S from P.

Exercise 6.18. Let K be a compact and convex subset of $\mathbb{R}^n$ and C a closed convex cone in $\mathbb{R}^n$. Prove that

$$\mathrm{rec}(K + C) = C.$$

Exercise 6.19. A half line in $\mathbb{R}^n$ is a subset of the form $\{x + \lambda d \,|\, \lambda \geq 0\}$, where $d \in \mathbb{R}^n \setminus \{0\}$ and $x \in \mathbb{R}^n$. Let C be a closed convex subset in $\mathbb{R}^n$. Prove that C contains a half line if and only if $\mathrm{rec}(C) \neq 0$. Does this hold without the assumption that C is closed?

Exercise 6.20. Let $K \subseteq \mathbb{R}^d$ be a non-empty compact subset. Prove that $\partial K \neq \emptyset$.

Exercise 6.21. Let $C \subseteq \mathbb{R}^d$ be a convex subset. Prove that the interior $\mathrm{int}(C) \subseteq C$ of C is a convex subset.

Exercise 6.22. Show how to avoid the use of Lemma 6.9 in the proof of Theorem 6.11 if C is a closed convex subset.

Exercise 6.23. Prove Gordan's theorem by applying Theorem 6.15 to the disjoint convex subsets

$$C_1 = \{Ax \,|\, x \in \mathbb{R}^n\}$$

and

$$C_2 = \{y \in \mathbb{R}^m \,|\, y < 0\}$$

in the notation of Theorem 4.17.

Exercise 6.24. A compact convex subset $C \subseteq \mathbb{R}^d$ with non-empty interior $\text{int}(C) \neq \emptyset$ is called a lattice free body if $\text{int}(C) \cap \mathbb{Z}^d = \emptyset$. Give a few examples of polyhedral and non-polyhedral lattice free bodies. Prove that a lattice free body not contained in a larger lattice free body (a so-called *maximal lattice free body*) is a polyhedron.

Chapter 7

Convex functions

Convex functions are special functions $f : C \to \mathbb{R}$ defined on a convex subset $C \subseteq \mathbb{R}^n$. They were introduced by the Danish mathematician Jensen (see §7.2) in [Jensen (1905)], where he writes

> *Jeg kan til Slutning ikke tilbageholde den Bemærkning at de foregaaende Undersøgelser synes at vise, at Begrebet »konveks Funktion« paa Grund af sin Simpelhed og elementære Karakter er nyttigt og vel egnet til at fortjene en Plads i Indledningen til de reelle Funktioners Theori næst efter et saa simpelt Begreb som »voksende Funktion«.*

This translates roughly into: "*Finally, I cannot withhold the remark that the preceding investigations seem to show that the concept »convex function« because of its simplicity and elementary character is useful and well suited to deserve a place in the introduction to the theory of real functions after a simple concept like »increasing function«.*" History has certainly proved Jensen right!

We first introduce convex functions of one variable by a motivating example leading to the famous inequality that inspired Jensen.

The inequality between the geometric and arithmetic means

You are approached by a questionable character asking for a loan of $\$1000$ over three years. You agree, but only for an interest of 30 % per year. He cannot pay any interest the first year. In the second year he promises to pay 5 % interest and in the third year he will pay an astounding 100 %. He claims that you are left with a loan that pays 35 % in interest per year, because

$$(0 + 5 + 100)/3 \% = 35 \%$$

he says. If you think about this for a while, you will almost surely discover that something is wrong. The average interest is computed not using the usual arithmetic mean but the so-called geometric mean. With the geometric mean we get

$$\sqrt[3]{1 \cdot 1.05 \cdot 2} = 1.28$$

telling you that the average interest is only 28 % per year.

However strange it may seem, this example is a good reason to learn about convex functions! One of the nice consequences of this theory is that

$$\sqrt[3]{abc} \leq \frac{a+b+c}{3}$$

for non-negative numbers a, b, c, where $=$ only occurs if $a = b = c$. Independently of the interest rates for the three year loan, you will always get cheated if you accept the "easy" way of computing the average interest unless the interest is constant.

Let us tell the story of the arithmetic and geometric mean. It is not too hard to prove that (see Exercise 7.1 for this and some fascinating mathematics involving fast computation of the digits of π)

$$\sqrt{ab} \leq \frac{a+b}{2} \tag{7.1}$$

for numbers $a, b \geq 0$ with equality if and only if $a = b$.

The inequality (7.1) is the first non-trivial case of the famous *inequality between the geometric and the arithmetic mean*. For three numbers $a, b, c \geq 0$ this inequality reads

$$\sqrt[3]{abc} \leq \frac{a+b+c}{3} \tag{7.2}$$

as we saw in the example. If we try to prove this as for two numbers, we expand (7.2) into

$$27abc \leq a^3 + 3ba^2 + 3ca^2 + 3b^2a + 3c^2a + 6bca + b^3 + c^3 + 3bc^2 + 3b^2c,$$

which is not so easy! We clearly need a genuine idea to prove the general case

$$\sqrt[n]{a_1 a_2 \cdots a_n} \leq \frac{a_1 + a_2 + \cdots + a_n}{n} \tag{7.3}$$

of the arithmetic geometric inequality. Even though (7.3) predates and inspired the theory of convex functions, we will view it in this more modern context.

7.1 Basics

Consider a function $f : C \to \mathbb{R}$, where C is a subset of $\mathbb{R}^n$. The graph of f is the subset of $\mathbb{R}^{n+1}$ given by $\{(x, y) \mid x \in C, y \in \mathbb{R}, \text{ and } y = f(x)\}$. The *epigraph* of f is the subset

$$\mathrm{epi}(f) = \{(x, y) \mid x \in C, \ y \in \mathbb{R} \text{ and } y \geq f(x)\}$$

of $\mathbb{R}^{n+1}$.

Figure 7.1: The epigraph of $f(x) = x^2$ is the set of points above the graph of f.

Definition 7.1. Let $C \subseteq \mathbb{R}^n$ be a convex subset. A *convex function* is a function $f : C \to \mathbb{R}$ such that

$$f((1 - \lambda)x + \lambda y) \leq (1 - \lambda)f(x) + \lambda f(y) \tag{7.4}$$

and a *concave function* is a function $f : C \to \mathbb{R}$, such that

$$f((1 - \lambda)x + \lambda y) \geq (1 - \lambda)f(x) + \lambda f(y)$$

for every $x, y \in C$ and every $\lambda \in \mathbb{R}$ with $0 \leq \lambda \leq 1$.

A convex (concave) function $f : C \to \mathbb{R}$ is called *strictly convex* (*strictly concave*) if

$$f((1 - \lambda)x + \lambda y) = (1 - \lambda)f(x) + \lambda f(y)$$

implies that $x = y$, for every $x, y \in C$ and every $\lambda \in \mathbb{R}$ with $0 < \lambda < 1$.

Notice that a strictly convex function satisfies $f((1 - \lambda)x + \lambda y) < (1 - \lambda)f(x) + \lambda f(y)$ for $x \neq y$ and $0 < \lambda < 1$ and that $-f$ is (strictly) concave if and only if f is (strictly) convex.

Our main emphasis will be on convex functions. The convexity of a function has a rather simple interpretation in terms of its epigraph.

Lemma 7.2. *Let $C \subseteq \mathbb{R}^n$ be a convex subset. A function $f : C \to \mathbb{R}$ is convex if and only epi(f) is a convex subset of $\mathbb{R}^{n+1}$.*

Proof.   Suppose epi(f) is a convex subset. For $(x, f(x)), (y, f(y)) \in \text{epi}(f)$ we have

$$(1 - \lambda)(x, f(x)) + \lambda(y, f(y))$$
$$= ((1 - \lambda)x + \lambda y, (1 - \lambda)f(x) + \lambda f(y)) \in \text{epi}(f),$$

for every $0 \le \lambda \le 1$ and $ux, y \in C$. Therefore

$$f((1 - \lambda)x + \lambda y) \le (1 - \lambda)f(x) + \lambda f(y)$$

and f is convex. If $(x_0, y_0), (x_1, y_1) \in \text{epi}(f)$, then $f(x_0) \le y_0$ and $f(x_1) \le y_1$. If f is convex, this implies

$$f((1 - \lambda)x_0 + \lambda x_1) \le (1 - \lambda)f(x_0) + \lambda f(x_1) \le (1 - \lambda)y_0 + \lambda y_1$$

for $0 \le \lambda \le 1$, so that

$$(1 - \lambda)(x_0, y_0) + \lambda(x_1, y_1) \in \text{epi}(f)$$

proving that epi(f) is a convex subset of $\mathbb{R}^{n+1}$.   □

The following example illustrates algebraic manipulations in proving convexity of a function from first principles.

Example 7.3. Let us consider the function $f(x) = ax + b$ from $\mathbb{R}$ to $\mathbb{R}$, where $a, b \in \mathbb{R}$ and check explicitly that f is a convex function. We must verify that $f((1 - \lambda)x + \lambda y) \le (1 - \lambda)f(x) + \lambda f(y)$ for every $0 \le \lambda \le 1$ and $x, y \in \mathbb{R}$:

$$f((1 - \lambda)x + \lambda y)$$
$$= a((1 - \lambda)x + \lambda y) + b$$
$$= a(1 - \lambda)x + (1 - \lambda)b + a\lambda y + \lambda b$$
$$= (1 - \lambda)(ax + b) + \lambda(ay + b)$$
$$= (1 - \lambda)f(x) + \lambda f(y)$$

proving that $f : \mathbb{R} \to \mathbb{R}$ is a convex function. You can easily check (with the same proof) that f is also a concave function. However, f is neither strictly convex nor strictly concave.

Consider $f(x) = x^2$ as a function from $\mathbb{R}$ to $\mathbb{R}$. From Lemma 7.2 and Figure 7.1 it seems that f is a convex function. To prove this precisely we need to show that

$$((1 - \lambda)x + \lambda y)^2 \leq (1 - \lambda)x^2 + \lambda y^2$$

for $0 \leq \lambda \leq 1$ and $x, y \in \mathbb{R}$. By a small computation one gets

$$(1 - \lambda)x^2 + \lambda y^2 - ((1 - \lambda)x + \lambda y)^2 = \lambda(1 - \lambda)(x^2 + y^2 - 2xy)$$
$$= \lambda(1 - \lambda)(x - y)^2.$$

From here it immediately follows that $f(x) = x^2$ is a strictly convex (and therefore convex) function.

However $f(x) = x^3$ is not a convex function from $\mathbb{R}$ to $\mathbb{R}$. This can be seen (formally) by putting $x = -1$, $y = 0$ and $\lambda = \frac{1}{2}$ in (7.4) noticing that

$$f((1 - \lambda)x + \lambda y) = -\tfrac{1}{8} > (1 - \lambda)f(x) + \lambda f(y) = -\tfrac{1}{2}.$$

Of course, sketching the epigraph of x^3 on $[-1, 0]$ also reveals its non-convexity.

7.2 Jensen's inequality

The Danish mathematician *Johan Ludvig William Valdemar Jensen* discovered a very interesting inequality around 1905. It appeared in a famous paper in the renowned journal *Acta Mathematica* in 1906. Today it bears the name *Jensen's inequality* and fits in retrospect perfectly in our theory of convex functions. It can be seen in the general context formulated below.

Corollary 7.4. *Let $\lambda_1, \ldots, \lambda_m \geq 0$ be real numbers with $\lambda_1 + \cdots + \lambda_m = 1$, where $m \geq 2$. If $C \subseteq \mathbb{R}^n$ is a convex subset and $f : C \to \mathbb{R}$ a convex function, then*

$$f(\lambda_1 x_1 + \cdots + \lambda_m x_m) \leq \lambda_1 f(x_1) + \cdots + \lambda_m f(x_m),$$

for $x_1, \ldots, x_m \in C$. If f is a strictly convex function and $\lambda_1, \ldots, \lambda_m > 0$, then

$$f(\lambda_1 x_1 + \cdots + \lambda_m x_m) = \lambda_1 f(x_1) + \cdots + \lambda_m f(x_m)$$

implies that $x_1 = \cdots = x_m$.

Proof.   The first claim is proved using induction on m. For $m = 2$ the claims are the definitions of a function being convex and strictly convex. For $m > 2$ we have for $\lambda = \lambda_2 + \cdots + \lambda_m = 1 - \lambda_1 > 0$,

$$
f(\lambda_1 x_1 + \cdots + \lambda_m x_m) = f\left(\lambda_1 x_1 + \lambda\left(\frac{\lambda_2}{\lambda}x_2 + \cdots + \frac{\lambda_m}{\lambda}x_m\right)\right)
$$

$$
\leq \lambda_1 f(x_1) + \lambda f\left(\frac{\lambda_2}{\lambda}x_2 + \cdots + \frac{\lambda_m}{\lambda}x_m\right) \tag{7.5}
$$

$$
\leq \lambda_1 f(x_1) + \lambda\left(\frac{\lambda_2}{\lambda}f(x_2) + \cdots + \frac{\lambda_m}{\lambda}f(x_m)\right) \tag{7.6}
$$

$$
= \lambda_1 f(x_1) + \cdots + \lambda_m f(x_m)
$$

where the convexity of f is used in deducing (7.5) and the induction assumption is used in deducing (7.6). In the proof of the second claim for a strictly convex function, the above inequalities are all equalities by assumption. Therefore $x_2 = \cdots = x_m$ by induction and

$$
x_1 = \frac{\lambda_2}{\lambda}x_2 + \cdots + \frac{\lambda_m}{\lambda}x_m = x_2 = \cdots = x_m
$$

by the definition of a strictly convex function. $\square$

Jensen never acquired a formal degree or held an academic appointment. He worked as a telephone engineer in Copenhagen doing mathematics in his spare time. He seems to have pursued mathematics for its beauty and his own enjoyment. Jensen appears on the cover of this book, not only for his contributions to convexity, but also as a tribute to the joyful and genuine pursuit of mathematics.

In [Jensen (1906)], Jensen introduced and studied midpoint convexity. A function $f : [a, b] \to \mathbb{R}$ is called *midpoint convex* if

$$
f\left(\frac{x + y}{2}\right) \leq \frac{f(x) + f(y)}{2}
$$

for every $x, y \in [a, b]$. In fact this was the definition of a convex function in [Jensen (1906)]. Jensen proved that a continuous midpoint convex function is convex in the sense of (7.4). This was inspired by a classical proof due to Cauchy of (7.3) — see Exercise 7.2. There exist midpoint convex functions, which are not convex in the sense of (7.4). Their existence has to do with the foundations of mathematics. They have never been observed in real life.

We will use Corollary 7.4 to prove the famous inequality (7.3) between the geometric and arithmetic mean. For this we assume that the exponential

$$\varphi\left(\frac{\Sigma a_\nu x_\nu}{\Sigma a_\nu}\right) \le \frac{\Sigma a_\nu \varphi(x_\nu)}{\Sigma a_\nu}$$

Figure 7.2: Jensen's inequality on postage stamp from Department of Mathematical Sciences, University of Copenhagen.

function $f(x) = e^x$ is a strictly convex function from $\mathbb{R}$ to $\mathbb{R}$ (this is proved later in Corollary 7.20). Let $\log : \{x \in \mathbb{R} \mid x > 0\} \to \mathbb{R}$ denote the natural logarithm i.e., the inverse function to f so that $f(\log(x)) = e^{\log(x)} = x$ for $x > 0$. For positive numbers $a_1, \ldots, a_n$ we have the identity

$$\begin{aligned}
\sqrt[n]{a_1 \cdots a_n} &= (a_1 \cdots a_n)^{1/n} = a_1^{1/n} \cdots a_n^{1/n} = e^{\log(a_1)/n} \cdots e^{\log(a_n)/n} \\
&= e^{1/n \log(a_1) + \cdots + 1/n \log(a_n)} \le \frac{1}{n} e^{\log(a_1)} + \cdots + \frac{1}{n} e^{\log(a_n)} \\
&= \frac{a_1 + \cdots + a_n}{n},
\end{aligned}$$

where the inequality above follows from Corollary 7.4 applied to $f(x) = e^x$. Corollary 7.4 also implies that equality occurs in (7.3) for $a_1, \ldots, a_n > 0$ if and only if $a_1 = \cdots = a_n$. Notice that we have used the fundamental identity $e^{x+y} = e^x e^y$ for $x, y \in \mathbb{R}$.

7.3 Minima of convex functions

Definition 7.5. Consider a function

$$f : S \to \mathbb{R},$$

where S is any subset of $\mathbb{R}^n$. A *global minimum* for f is a point $x_0 \in S$, such that

$$f(x_0) \le f(x) \tag{7.7}$$

for every $x \in S$. A point $x_0 \in S$ is called a *local minimum* if (7.7) only holds locally i.e., if there exists $\epsilon > 0$, such that $f(x_0) \le f(x)$ for $x \in S$ with $|x - x_0| < \epsilon$. *Global and local maxima* are defined similarly by reversing the inequality in (7.7). A global minimum or maximum for f is called a *global extremum* for f. A local minimum or maximum for f is called a *local extremum*.

Figure 7.3: A local but not global minimum.

A global minimum is clearly a local minimum, but not the other way around. For convex functions, we have the following remarkable result.

Lemma 7.6. *Let $f : C \to \mathbb{R}$ be a convex function. If $x_0 \in C$ is a local minimum, then x_0 is a global minimum. If f is strictly convex, then a global minimum for f is unique.*

Proof. By definition there exists $\epsilon > 0$, such that $f(x_0) \leq f(x)$, when $x \in C$ and $|x - x_0| < \epsilon$. Suppose that x_0 is not a global minimum. Then there exists $x_1 \in C$ with $f(x_1) < f(x_0)$. Consider the point

$$x_t = (1 - t)x_0 + tx_1 \in C,$$

where $0 < t < 1$. Then

$$f(x_t) \leq (1 - t)f(x_0) + tf(x_1) < (1 - t)f(x_0) + tf(x_0) = f(x_0).$$

Since $|x_t - x_0| = t|x_1 - x_0|$, we can choose $t > 0$ sufficiently small such that $|x_t - x_0| < \epsilon$ implying $f(x_0) \leq f(x_t)$, since x_0 is a local minimum. This contradicts that $f(x_t) < f(x_0)$ for every $0 < t < 1$. Let f be strictly convex and let x_0 be a global minimum for f. If $x_1 \in C$, $x_1 \neq x_0$ and $f(x_1) = f(x_0)$, then

$$f((1 - \lambda)x_0 + \lambda x_1) < (1 - \lambda)f(x_0) + \lambda f(x_1) = f(x_0)$$

for $0 < \lambda < 1$. This would contradict that global minimality of x_0, since $x_0 \neq (1 - \lambda)x_0 + \lambda x_1 \in C$ for $0 < \lambda < 1$. $\square$

7.4 Convex functions of one variable

For the rest of this chapter we shall study convex functions of one variable i.e., convex functions $f : C \to \mathbb{R}$, where $C \subseteq \mathbb{R}$ is an interval. We begin by showing that convex functions are continuous on open intervals by tweaking (7.4).

Theorem 7.7. *A convex function $f : [a, b] \to \mathbb{R}$ is continuous on the open interval (a, b).*

Proof. We will prove that f is continuous at a point $x_0 \in (a, b)$. The convex function $g(x) = f(x + x_0) - f(x_0)$ defined on $(a - x_0, b - x_0)$ is continuous at 0 if and only if f is continuous at x_0. We may therefore assume that $x_0 = 0 \in (a, b)$ and $f(0) = 0$. For a sufficiently small $\delta > 0$, $I = [-\delta, \delta] \subseteq (a, b)$. If $x \in I$, then

$$f(x) = f(|x| \operatorname{sgn}(x)) = f\left(\frac{|x|}{\delta} \operatorname{sgn}(x)\delta\right)$$
$$= f\left(\frac{|x|}{\delta} \operatorname{sgn}(x)\delta + \left(1 - \frac{|x|}{\delta}\right) \cdot 0\right)$$
$$\leq \frac{|x|}{\delta} f(\operatorname{sgn}(x)\delta)$$

by (7.4) using that $f(0) = 0$. Putting $M = \max(f(-\delta), f(\delta))$ we therefore get $f(x) \leq \frac{M}{\delta}|x|$ for $x \in I$. Now notice that $-f(-x) \leq f(x)$ for $x \in I$ (see Exercise 7.5). Since $-I = I$ we deduce, for $x \in I$ that

$$|f(x)| \leq \frac{M}{\delta}|x|. \tag{7.8}$$

If (x_n) is a sequence in (a, b) converging to 0, then there exists $N \in \mathbb{N}$ such that $x_n \in I$ for $n \geq N$. Now (7.8) forces $(f(x_n))$ to converge to $f(0) = 0$. This shows that f is continuous at 0. $\qquad\square$

The above continuity proof is quite formal and not very geometric. It has, however, the advantage that it generalizes naturally to higher dimensions (see Theorem 10.17), where the intuition presented in the next important result is absent.

Lemma 7.8. *Let $f : [a, b] \to \mathbb{R}$ be a convex function. Then*

$$\frac{f(x) - f(a)}{x - a} \leq \frac{f(b) - f(a)}{b - a} \leq \frac{f(b) - f(x)}{b - x} \tag{7.9}$$

for $a < x < b$.

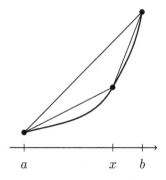

Figure 7.4: Visualization of (7.9).

Proof.   The convex combination

$$x = a + \frac{x-a}{b-a}(b-a) = \left(1 - \frac{x-a}{b-a}\right)a + \frac{x-a}{b-a}b$$

of a and b shows that

$$f(x) \le \left(1 - \frac{x-a}{b-a}\right)f(a) + \frac{x-a}{b-a}f(b),$$

which gives $(f(x)-f(a))(b-a) \le (x-a)(f(b)-f(a))$ and the first inequality follows. The same convex combination written as

$$x = b + \frac{x-b}{b-a}(b-a) = \frac{b-x}{b-a}a + \left(1 - \frac{b-x}{b-a}\right)b$$

gives the other inequality.   □

7.5   Differentiable functions of one variable

Before turning to differentiable convex functions of one variable we will recall a few basic notions on differentiable functions. You have probably already seen this material in a previous course. Consider the following pages an opportunity to review the concepts.

It is rewarding to study the writings of people known for their intuitive and playful approach. On and off I find myself browsing *The Feynman Lectures on Physics* when I am fed up with terse and rigid textbooks. Here is a very illuminating story that Feynman[1] used to illustrate the concept of differentiability (see §8–2 in volume I of [Feynman *et al.* (2006)]).

---

[1]Richard P. Feynman (1918–1988). American physicist.

At the point where the lady in the car is caught by a cop, the cop comes up to her and says. "Lady, you were going 60 miles an hour!" She says "That's impossible, sir, I was travelling for only seven minutes. It is ridiculous – how can I go 60 miles an hour when I wasn't going an hour".

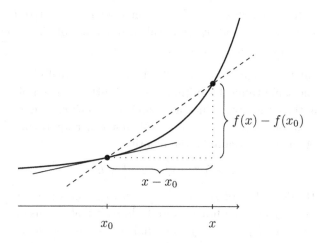

Figure 7.5: The slope of the solid curve at x_0 is the limit (7.10) as x tends to x_0.

The lady asks a very good question and a satisfactory answer is not so immediate. We will briefly recall the basics of differentiable functions of one variable.

The derivative of a function at a point x_0 is a subtle concept. The remarkable idea is to investigate the slope of the chord between $(x_0, f(x_0))$ and $(x, f(x))$ as x approaches x_0 i.e.,

$$\lim_{x \to x_0} \frac{f(x) - f(x_0)}{x - x_0}. \tag{7.10}$$

Several hundred years ago, Newton defined the derivative as the slope just before $\Delta x = x - x_0$ becomes zero. In modern terms, this can be phrased as follows: f is differentiable at x_0 if there exists a number $c \in \mathbb{R}$, such that for every $\epsilon > 0$ there exists δ with

$$\left| \frac{f(x) - f(x_0)}{x - x_0} - c \right| < \epsilon \tag{7.11}$$

for $|x - x_0| < \delta$. Our working definition of differentiability will be the following.

Definition 7.9. A function $f : (a, b) \to \mathbb{R}$ is *differentiable at* x_0 if for some $\delta > 0$ there exists a function $\epsilon : (-\delta, \delta) \to \mathbb{R}$ continuous in 0 with $\epsilon(0) = 0$ and a number c, such that

$$f(x_0 + h) - f(x_0) = ch + \epsilon(h)h \qquad (7.12)$$

for $x_0 + h \in (a, b)$ and $h \in (-\delta, \delta)$. The number c is denoted $f'(x_0)$ and called *the derivative* of f at x_0; f is called differentiable if its derivative exists at every $x_0 \in (a, b)$.

You should at this point check that (7.11) and (7.12) really are two ways of defining the same thing. Definition 7.9 is admittedly a complicated way of perceiving the derivative, which after all is of geometric origin. One may view the definition as explaining the best first order approximation to f around x_0. As a simple example, the identity

$$(x_0 + h)^2 - x_0^2 = 2x_0 h + h^2$$

shows that the derivative of $f(x) = x^2$ at x_0 is $2x_0$. Here $\epsilon(h) = h$ in (7.12).

We may view $f'(x)$ as a function in itself, so it makes sense to consider the *second derivative* $f''(x) = (f')'(x)$ and so on. For the n-th derivative we use the notation $f^{(n)}(x)$. A function defined on an open interval (a, b) is called n times differentiable if its *n-th derivative* exists on (a, b).

7.5.1   The Newton-Raphson method for finding roots

Consider for $a \in \mathbb{R}$ the sequence given by $x_0 = 1$ and

$$x_{n+1} = \frac{1}{2} \left(x_n + \frac{a}{x_n} \right). \qquad (7.13)$$

For $a = 1$, the sequence is constant $1 = x_0 = x_1 = \cdots$. For $a = 2$ a calculator gives the first few numbers approximately

$$x_0 = 1, \quad x_1 = 1.5, \quad x_2 = 1.416\,666\,7, \quad x_3 = 1.414\,22,$$
$$x_4 = 1.414\,215\,7, \quad x_5 = 1.414\,213\,6, \quad x_6 = 1.4142136.$$

This sequence certainly looks convergent and the limit seems to be $\sqrt{2}$. You may be surprised to learn that this is related to (7.12): suppose that $f : \mathbb{R} \to \mathbb{R}$ is a differentiable function and that $f(\xi) = 0$. We wish to compute an approximation to ξ. Suppose that x_0 is close to ξ. Then (7.12) tells us that

$$f(\xi) - f(x_0) \quad \text{is close to} \quad f'(x_0)(\xi - x_0). \qquad (7.14)$$

Using $f(\xi) = 0$ and regarding "close to" as $=$ in (7.14), we compute

$$x_1 = x_0 - \frac{f(x_0)}{f'(x_0)}$$

provided that $f'(x_0) \neq 0$. In favorable cases, x_1 is closer to ξ than x_0. This procedure may be repeated with x_1 instead of x_0 provided that $f'(x_1) \neq 0$. If the derivatives along the way are $\neq 0$ we end up with a sequence $x_0, x_1, x_2, \ldots$ — in many cases this sequence converges to ξ. This iterative method (see Figure 7.6) is called the *Newton-Raphson*[2] *method*. Putting $f(x) = x^2 - a$ one recovers the iterative formula (7.13).

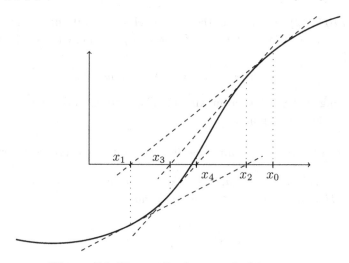

Figure 7.6: Newton-Raphson method in action.

This algorithm is one of the many reasons that the whole concept of differentiability is deep and somewhat miraculous. The method generalizes to differentiable functions of several variables (see §8.1.1).

7.5.2 *Critical points and extrema*

The derivative can be used in finding local extrema for a differentiable function.

---

[2]Joseph Raphson (1648–1715). English mathematician.

Definition 7.10. A point $x_0 \in (a,b)$ is called a *critical point* for a differentiable function $f : (a,b) \to \mathbb{R}$ if

$$f'(x_0) = 0.$$

Lemma 7.11. *Let* $f : (a,b) \to \mathbb{R}$ *be a differentiable function. If* $\xi \in (a,b)$ *is a local extremum, then* ξ *is a critical point for* f.

Proof. Suppose that ξ is a local maximum and that

$$f(\xi + h) - f(\xi) = f'(\xi)h + \epsilon(h)h$$

according to (7.12). If $f'(\xi) > 0$, then we can choose $\delta > 0$ sufficiently small, such that $|\epsilon(h)| < f'(x_0)$ if $0 \le h < \delta$, since $\epsilon(0) = 0$ and ϵ is continuous in 0. Therefore

$$f(\xi + h) - f(\xi) = (f'(\xi) + \epsilon(h))h > 0,$$

contradicting that ξ is a local maximum. The proof is similar for $f'(\xi) < 0$ and if ξ is a local minimum. $\square$

We will need the famous *mean value theorem* stated and proved below.

Theorem 7.12. *Let* $f : [a,b] \to \mathbb{R}$ *be continuous and differentiable on* (a,b). *Then there exists* $\xi \in (a,b)$ *such that*

$$f'(\xi) = \frac{f(b) - f(a)}{b - a}. \tag{7.15}$$

Proof. Consider the function

$$g(x) = \frac{f(b) - f(a)}{b - a}(x - a) - f(x) + f(a).$$

Then $g : [a,b] \to \mathbb{R}$ is continuous and differentiable on (a,b) with $g(a) = g(b) = 0$. If $g(x) = 0$ for every $x \in (a,b)$, then any $\xi \in (a,b)$ satisfies (7.15). Recall that g being continuous on a closed interval $[a,b]$ attains its maximum and minimum values for some $x \in [a,b]$ (see §A.7). So if $g(x)$ is not constant $(= 0)$, there must exist a local extremum $\xi \in (a,b)$ and

$$g'(\xi) = \frac{f(b) - f(a)}{b - a} - f'(\xi) = 0$$

by Lemma 7.11. Now the result follows. $\square$

The mean value theorem is extremely useful. Before giving the first application we insert a definition.

Definition 7.13. A function $f : S \to \mathbb{R}$ with $S \subseteq \mathbb{R}$ is called *increasing* if

$$x \leq y \Rightarrow f(x) \leq f(y)$$

and *strictly increasing* if

$$x < y \Rightarrow f(x) < f(y)$$

for $x, y \in S$.

Corollary 7.14. *Let $f : (a, b) \to \mathbb{R}$ be a differentiable function. Then f is increasing if and only if $f'(x) \geq 0$ for every $x \in (a, b)$. If $f'(x) > 0$ for every $x \in (a, b)$, then f is strictly increasing.*

Proof. Assume that $f'(x) \geq 0$ for every $x \in (a, b)$ and suppose that there exist $x_1, x_2 \in (a, b)$ with $x_1 < x_2$ and $f(x_1) > f(x_2)$. Then Theorem 7.12 implies the existence of $\xi \in (x_1, x_2)$, such that

$$f'(\xi) = \frac{f(x_2) - f(x_1)}{x_2 - x_1} < 0$$

contradicting our assumption that $f'(x) \geq 0$ for every $x \in (a, b)$. A similar proof shows that f is strictly increasing if $f'(x) > 0$ for $x \in (a, b)$. On the other hand if f is increasing, then

$$\frac{f(x + h) - f(x)}{h} \geq 0$$

for every $x \in (a, b)$ and $h > 0$ with $x + h < b$. This shows that $f'(x) \geq 0$ for every $x \in (a, b)$. $\qquad\square$

If z_0 is a critical point for a differentiable function f we cannot conclude that z_0 is a local extremum for f (example?). If f' is differentiable, the second derivative $f''(z_0)$ contains further information about the critical point. In fact, one of the wonders of (infinitely) differentiable functions is that they are in a certain sense uniquely determined from the numbers

$$f(x_0), \quad f'(x_0), \quad f''(x_0), \quad \ldots,$$

where x_0 is point, where f is defined. To make this more precise we need to introduce Taylor polynomials.

7.6   Taylor polynomials

Let us for a moment put rigor aside. Suppose that the infinite expression

$$a_0 + a_1 x + a_2 x^2 + a_3 x^3 + \cdots \tag{7.16}$$

defines a function $f : \mathbb{R} \to \mathbb{R}$. Can we recover $a_0, a_1, \ldots$ from f? Clearly $a_0 = f(0)$. Term wise differentiation of (7.16) gives

$$f'(x) = a_1 + 2a_2 x + 3a_3 x^2 + \cdots$$

and $a_1 = f'(0)$. Similarly

$$f''(x) = 2a_2 + 6a_3 x + \cdots$$

and $a_2 = f''(0)/2$. In general we have

$$a_n = \frac{f^{(n)}(0)}{n!}. \tag{7.17}$$

This argument of undetermined coefficients goes back to Maclaurin[3]. It is not precise in the sense of modern mathematics, but creative imprecision often leads to new ideas. If you apply the method above to $f(x) = e^x$ (here $f^{(n)}(x) = e^x$) you get

$$e^x = 1 + x + \frac{x^2}{2} + \frac{x^3}{6} + \cdots + \frac{x^n}{n!} + \cdots \tag{7.18}$$

We will try and make this precise. If

$$f(x) = a_0 + a_1 x + \cdots + a_n x^n$$

is a polynomial of degree n ($a_n \neq 0$), then certainly the above argument is precise and we get

$$f(x) = f(0) + f'(0)x + \frac{f''(0)}{2}x^2 + \cdots + \frac{f^{(n)}(0)}{n!}x^n.$$

Inspired by Maclaurin's idea we introduce the Taylor[4] polynomial.

Definition 7.15. Let $f : [a, b] \to \mathbb{R}$ be an n times differentiable function on (a, b). Then the *Taylor polynomial*, P_{n,x_0}, of order n for f at $x_0 \in (a, b)$ is

$$f(x_0) + f'(x_0)(x - x_0) + \frac{f''(x_0)}{2}(x - x_0)^2 + \cdots + \frac{f^{(n)}(x_0)}{n!}(x - x_0)^n.$$

[3]Colin Maclaurin (1698–1746). Scottish mathematician.
[4]Brook Taylor (1685–1731). English mathematician.

Notice that $P = P_{n,x_0}$ satisfies

$$P(x_0) = f(x_0), \quad P'(x_0) = f'(x_0), \quad \ldots, \quad P^{(n)}(x_0) = f^{(n)}(x_0).$$

To make a formula like (7.18) precise, it is important to know how well the Taylor polynomial P_{n,x_0} approximates f around x_0.

Theorem 7.16. *Let* $f : [a, b] \to \mathbb{R}$ *be an* $n+1$ *times differentiable function on* (a, b). *If* $x, x_0 \in (a, b)$ *with* $x_0 \neq x$, *then there exists* η *between* x_0 *and* x, *such that*

$$f(x) - P_{n,x_0}(x) = \frac{f^{(n+1)}(\eta)}{(n+1)!}(x - x_0)^{n+1}. \tag{7.19}$$

Proof. Suppose that the constant K is determined by

$$f(x) - P_{n,x_0}(x) = K(x - x_0)^{n+1}.$$

Consider the function

$$\varphi(t) = f(x) - P_{n,t}(x) - K(x - t)^{n+1},$$

where x_0 is replaced by t. Then an enlightening explicit computation (do it!) shows that

$$\varphi'(t) = -\frac{f^{(n+1)}(t)}{n!}(x - t)^n + K(n+1)(x - t)^n. \tag{7.20}$$

Since $\varphi(x) = \varphi(x_0) = 0$, there exists η between x_0 and x, such that $\varphi'(\eta) = 0$ by Theorem 7.12. Substituting η for t in (7.20) and solving $\varphi'(\eta) = 0$ for K, gives (7.19). $\qquad \square$

The base of the natural logarithm

$$e = 2.718\,281\,828\,459\,05\ldots \tag{7.21}$$

is one of the fundamental mathematical constants. Putting $f(x) = e^x$, we know that $f'(x) = e^x$ and we can use Theorem 7.16 with $x_0 = 0$ and $x = 1$ to deduce that

$$e - \left(1 + 1 + \frac{1}{2!} + \frac{1}{3!} + \cdots + \frac{1}{n!}\right) < \frac{e}{(n+1)!} < \frac{1}{n!} \tag{7.22}$$

for $n > 2$. Approximating e by the 10-th Taylor polynomial you therefore get an error of at most $1/10! = 1/3\,628\,800 = 0.000\,000\,276\ldots$. The accuracy in (7.21) is attained going to the 17-th Taylor polynomial. If you want to dig deeper into pure mathematics, you can even use (7.22) to prove that e is not a rational number.

Example 7.17. Let us use Taylor polynomials to compute an approximation to $\sqrt[3]{1007}$. Here we let $f(x) = \sqrt[3]{1000 + x}$ and $x_0 = 0$. Explicit differentiation gives

$$f'(x) = \tfrac{1}{3}(1000 + x)^{-\frac{2}{3}}$$
$$f''(x) = -\tfrac{2}{9}(1000 + x)^{-\frac{5}{3}}$$
$$f'''(x) = \tfrac{10}{27}(1000 + x)^{-\frac{8}{3}}.$$

The quadratic Taylor polynomial $P_{2,0}(x)$ for f is

$$10 + \frac{1}{300}x - \frac{1}{900\,000}x^2.$$

For $x = 7$ we have

$$P_{2,0}(7) = \frac{9\,020\,951}{900\,000},$$

which is

$$10.023\,278\,888\,88\ldots.$$

This is a pretty good approximation to $\sqrt[3]{1007}$. In fact, the error term from Theorem 7.16 is

$$\frac{f'''(\eta)}{3!}7^3 = \frac{10}{6 \cdot 27}\eta^{-\frac{8}{3}}7^3 < \frac{1}{10}\frac{1}{10^8}10^3 = 10^{-6},$$

where $1000 < \eta < 1007$. I have to break down and admit that in this digital age, a slightly sophisticated computer program gives

$$\sqrt[3]{1007} = 10.023\,279\,099\,634\,3\ldots.$$

Then again, Taylor polynomials have had lasting importance for three centuries. Will your favorite computer program be selling fifty years from now?

We conclude this section with an important application of the higher derivatives enabling you to identify minima and maxima at critical points.

Theorem 7.18. *Let x_0 be a critical point of an $n + 1$ times differentiable function $f : (a, b) \to \mathbb{R}$ with $f^{(n+1)}(x)$ a continuous function,*

$$f''(x_0) = \cdots = f^{(n-1)}(x_0) = 0$$

and $f^{(n)}(x_0) \neq 0$. If n is even, then x_0 is a local minimum if $f^{(n)}(x_0) > 0$ and a local maximum if $f^{(n)}(x_0) < 0$. If n is odd, then x_0 is not a local extremum.

Proof. Theorem 7.16 shows that

$$f(x_0 + h) - f(x_0) = \left(\frac{f^{(n)}(x_0)}{n!} + \frac{f^{(n+1)}(\eta)}{(n+1)!} h \right) h^n$$

for h numerically small and η between x_0 and $x_0 + h$. There exists M with $|f^{(n+1)}(x)| \leq M$ for every x in the closed interval with endpoints x_0 and $x_0 + h$ by the continuity of $f^{(n+1)}$. The proof is now similar to the proof of Lemma 7.11. $\qquad\square$

7.7 Differentiable convex functions

A convex function $f : [a, b] \to \mathbb{R}$ does not have to be differentiable on (a, b). Here the function $f(x) = |x|$ is a prominent example. This function is convex on $[-1, 1]$ but not differentiable at $x_0 = 0$.

The following characterization of convex differentiable functions is extremely useful.

Theorem 7.19. *Let $f : (a, b) \to \mathbb{R}$ be a differentiable function. Then f is convex if and only if f' is increasing. If f' is strictly increasing, then f is strictly convex.*

Proof. First suppose that f is convex and differentiable. If $a < x \leq y < b$, then

$$\frac{f(x) - f(x - h)}{h} \leq \frac{f(y + h) - f(y)}{h}$$

for sufficiently small $h > 0$ by using Lemma 7.8 twice. Letting $h \to 0$ this shows that

$$f'(x) \leq f'(y).$$

Therefore f' is increasing. Assume now that f' is increasing. We wish to prove that

$$f((1 - \lambda)x + \lambda y)) \leq (1 - \lambda)f(x) + \lambda f(y) \tag{7.23}$$

for $x, y \in (a, b)$ and $0 \leq \lambda \leq 1$. This is clear if $\lambda = 0$ or $\lambda = 1$. Assume that $0 < \lambda < 1$ and $x < y$. Then Theorem 7.12 gives the two identities

$$f((1 - \lambda)x + \lambda y) - f(x) = f'(\xi_1)\lambda(y - x)$$
$$f(y) - f((1 - \lambda)x + \lambda y) = f'(\xi_2)(1 - \lambda)(y - x),$$

where $x < \xi_1 < (1 - \lambda)x + \lambda y < \xi_2 < y$. Multiplying the first identity by $1 - \lambda$ and the second by λ and subtracting the first from the second and using the assumption $f'(\xi_1) \leq f'(\xi_2)$, we get

$$(1 - \lambda)(f((1 - \lambda)x + \lambda y) - f(x)) \leq \lambda(f(y) - f((1 - \lambda)x + \lambda y)) \quad (7.24)$$

proving (7.23). If f' is strictly increasing, then $\leq$ in (7.24) may be replaced by $<$ showing that f is strictly convex. $\qquad \square$

We will mostly encounter differentiable functions $f : (a, b) \to \mathbb{R}$ with the property that f'' exists (f is twice differentiable). In this case Corollary 7.14 implies the following.

Corollary 7.20. *Let $f : (a, b) \to \mathbb{R}$ be a twice differentiable function. Then f is convex if and only if $f''(x) \geq 0$ for every $x \in (a, b)$. If $f''(x) > 0$ for every $x \in (a, b)$, then f is strictly convex.*

Corollary 7.20 gives an instant proof of the fact that the exponential function $f(x) = e^x$ is strictly convex. In Example 7.3 we needed to invent an algebraic trick to prove that $f(x) = x^2$ is a strictly convex function. Now you can see this as an immediate consequence of Corollary 7.20.

Another important characterization of convex differentiable functions is given by the following.

Theorem 7.21. *Let $f : (a, b) \to \mathbb{R}$ be a differentiable function. Then f is convex if and only if*

$$f(y) \geq f(x) + f'(x)(y - x)$$

for every $x, y \in (a, b)$.

Proof.   Suppose that $x < x_1 < y$. Then Lemma 7.8 shows that

$$\frac{f(x_1) - f(x)}{x_1 - x} \leq \frac{f(y) - f(x)}{y - x}.$$

Letting x_1 converge to x from the right we end up with the desired inequality

$$f'(x) \leq \frac{f(y) - f(x)}{y - x}.$$

When $y < x$ the same proof works. On the other hand if

$$f(y) \geq f(x) + f'(x)(y - x) \qquad (7.25)$$

for every $x, y \in (a, b)$, then we will prove that $x < y$ implies $f'(x) \leq f'(y)$ and therefore that f is convex by Theorem 7.19. By (7.25) we have

$$f(y) \geq f(x) + f'(x)(y - x)$$
$$f(x) \geq f(y) + f'(y)(x - y) = f(y) - f'(y)(y - x),$$

which gives

$$f(x) + f'(y)(y - x) \geq f(x) + f'(x)(y - x).$$

Therefore $f'(x) \leq f'(y)$, since $x < y$. $\qquad\square$

Theorem 7.21 gives you yet another idea why convex differentiable functions are very special! In fact, it tells you that if $f : (a, b) \to \mathbb{R}$ is a convex differentiable function and $x_0 \in (a, b)$ satisfies $f'(x_0) = 0$, then

$$f(x_0) \leq f(x)$$

for every $x \in (a, b)$ i.e., x_0 is a global minimum!

7.8 Exercises

Exercise 7.1.

(i) Prove that

$$\sqrt{ab} \leq \frac{a + b}{2}$$

for $0 \leq a \leq b$ with equality if and only if $a = b$.

(ii) Prove that $a < \sqrt{ab}$ and $(a + b)/2 < b$ for $0 \leq a < b$.

(iii) Start with two numbers a and b with $0 \leq a \leq b$ and define

$$a_{n+1} = \sqrt{a_n b_n}$$
$$b_{n+1} = \frac{a_n + b_n}{2},$$

where $a_0 = a$ and $b_0 = b$. Prove for $n \geq 1$ that

$$b_n - a_n < \left(\tfrac{1}{2}\right)^n (b - a)$$

if $a \neq b$.

(iv) Let $s = \lim_{n \to \infty} a_n$ and $t = \lim_{n \to \infty} b_n$. Prove, after you have convinced yourself that these limits exist, that $s = t$.

Note. The common limit $M(a, b)$ of the two sequences (a_n) and (b_n) is called the *arithmetic-geometric mean*. The arithmetic-geometric mean provides fast converging approximations to π. A rapidly converging formula (already known to Gauss, who computed several decimals of π by hand) is (see Theorem 5 in [Almkvist and Berndt (1988)])

$$\pi = \frac{4M(1, \frac{1}{\sqrt{2}})^2}{1 - \sum_{n=1}^{\infty} 2^{n+1}(b_n^2 - a_n^2)},$$

where $a_0 = 1$ and $b_0 = 1/\sqrt{2}$ in the notation of c). Gauss discovered in 1799 the remarkable connection between the arithmetic-geometric mean and elliptic integrals of the first kind displayed in the formula

$$G = \frac{2}{\pi} \int_0^1 \frac{dx}{\sqrt{1 - x^4}},$$

where G is the reciprocal of the arithmetic-geometric mean of 1 and $\sqrt{2}$ (also called Gauss' constant).

Exercise 7.2. The following exercise aims at proving the inequality between the geometric and arithmetic means from scratch, without using Jensen's inequality. The proof presented here inspired Jensen's paper from 1906. It is due to Cauchy and goes back to 1821. All variables in this exercise denote non-negative real numbers.

(i) Let $a, b, c, d \geq 0$ be real numbers. Prove that

$$\sqrt[4]{abcd} \leq \frac{a + b + c + d}{4}$$

using $\sqrt[4]{x} = \sqrt{\sqrt{x}}$.

(ii) Prove by induction on m that

$$\sqrt[2^m]{x_1 \cdots x_{2^m}} \leq \frac{x_1 + \cdots + x_{2^m}}{2^m}$$

for $m \geq 1$.

(iii) Suppose now that $n \in \mathbb{N}$,

$$\alpha = \frac{x_1 + \cdots + x_n}{n}$$

and $m \in \mathbb{N}$ is chosen such that $2^m \geq n$. Show that

$$\alpha = \frac{x_1 + \cdots + x_n + (2^m - n)\alpha}{2^m} \geq \sqrt[2^m]{x_1 \cdots x_n \alpha^{2^m - n}}.$$

Use this to prove the inequality between the geometric and arithmetic means.

Exercise 7.3. Prove that

$$p(x,y) = x^2 y^2 (x^2 + y^2 - 3) + 1 \geq 0$$

for every $x, y \in \mathbb{R}$.

Hint: put $z^2 = 3 - x^2 - y^2$ for $x^2 + y^2 - 3 < 0$ and use the inequality between the geometric and arithmetic mean.

Note. The expression $p(x,y)$ is a polynomial in two variables x and y. One can prove that it is not a sum of squares of polynomials i.e., there does not exist finitely many polynomials $p_1(x,y), \ldots, p_n(x,y)$, such that

$$p(x,y) = p_1(x,y)^2 + \cdots + p_n(x,y)^2.$$

This polynomial was found by Motzkin in 1967 and is usually referred to as Motzkin's polynomial. Surprisingly $p(x,y)$ is a sum of squares of rational functions i.e., fractions $f(x,y)/g(x,y)$ of polynomials $f(x,y)$ and $g(x,y)$. This is related to Hilbert's[5] 17th problem, which was solved by Emil Artin[6] in 1927.

Exercise 7.4. Let $f : \mathbb{R} \to \mathbb{R}$ be a convex function. Prove that $f(ax + b)$ is a convex function for $a, b \in \mathbb{R}$.

Exercise 7.5. Let $f : \mathbb{R}^n \to \mathbb{R}$ be a convex function with $f(0) = 0$. Prove that

$$-f(x) \leq f(-x)$$

for every $x \in \mathbb{R}^n$. Show that $f(x) = 0$ if $f(x) \leq 0$ for every $x \in \mathbb{R}^n$.

Exercise 7.6. Is $|f(x)|$ a convex function if $f(x)$ is a convex function?

Exercise 7.7.

(i) Prove from scratch that $f(x) = x^4$ is a convex function only using that x^2 is a convex function.

(ii) Prove that $f(x) = x^4$ is a strictly convex function. Is $f(x) = x^6$ a strictly convex function?

---

[5]David Hilbert (1862–1943). German mathematician.
[6]Emil Artin (1898–1962). Austrian-American mathematician.

(iii) Prove that f is (strictly) convex if and only if $-f$ is (strictly) concave.

Exercise 7.8. Let $f : [a, b] \to \mathbb{R}$ be a concave function with $f(x) > 0$ for $x \in [a, b]$. Prove that $\log f(x)$ is a concave function.

Exercise 7.9. Let C denote the set of convex functions $f : \mathbb{R} \to \mathbb{R}$.

(i) Prove that $f + g, \lambda f \in C$ if $f, g \in C$ and $\lambda \geq 0$.
(ii) How is the set V of all functions $f : \mathbb{R} \to \mathbb{R}$ a vector space over the real numbers $\mathbb{R}$?
(iii) Show that C is a convex cone in V.
(iv) Is $1 + 23412x^{10} + 23x^{48}$ a convex function?

Exercise 7.10. Let $f : [a, b] \to \mathbb{R}$ be a convex function. Is f a continuous function on $[a, b]$?

Exercise 7.11. Sketch the graph of a differentiable function f, a zero ξ of f and a point x_0, such that the Newton-Raphson method fails to converge starting from x_0.

Exercise 7.12. Let $f : [a, b] \to \mathbb{R}$ be a twice differentiable function on (a, b) with $f'' \geq 0$. Is $f : [a, b] \to \mathbb{R}$ a convex function?

Exercise 7.13. Give an example of a strictly increasing differentiable function $f : [a, b] \to \mathbb{R}$ along with $\xi \in (a, b)$ such that $f'(\xi) = 0$.

Exercise 7.14. Give an example of a function $f : [-1, 1] \to \mathbb{R}$ differentiable on $(-1, 1)$ with $f'(0) = 0$, but where 0 is not a local extremum.

Exercise 7.15. Prove in detail that the function $f(x) = |x|$ fails to be differentiable at $x = 0$. Is f a convex function?

Exercise 7.16. In Babylonian mathematics (3000 B.C.) one encounters the formula
$$d = h + \frac{w^2}{2h}$$
for the length of the diagonal d in a rectangle with height h and width w. Give an example showing that the formula is wrong. When is it a reasonable approximation?

(Look at the Taylor expansion for $\sqrt{1 + x}$.)

Exercise 7.17. Show that $f : \{x \in \mathbb{R} \,|\, x \geq 0\} \to \mathbb{R}$ given by

$$f(x) = -\sqrt{x}$$

is a strictly convex function.

Exercise 7.18. Find a non-empty open subset $S \subseteq \mathbb{R}$, such that

$$f(x) = \sin(x)^2 : \mathbb{R} \to \mathbb{R}$$

is a convex function on S.

Exercise 7.19. Let $\varphi, f : \mathbb{R} \to \mathbb{R}$ be convex functions, where f is increasing. Prove that $f(\varphi(x))$ is a convex function.

Exercise 7.20. Give conditions on $a_0, a_1, a_2, a_3, a_4 \in \mathbb{R}$ ensuring that $a_0 + a_1 x + a_2 x^2 + a_3 x^3 + a_4 x^4$ is a convex function.

Exercise 7.21. Let $p, q > 0$ be real numbers with $1/p + 1/q = 1$. Prove that

$$ab \leq \frac{a^p}{p} + \frac{b^q}{q}$$

for real numbers $a, b \geq 0$. This is called *Young's inequality*.

Exercise 7.22. Prove that a bounded differentiable convex function $f : \mathbb{R} \to \mathbb{R}$ is constant (bounded means that there exists $M \in \mathbb{R}$, such that $|f(x)| \leq M$ for every $x \in \mathbb{R}$).

Exercise 7.23. Use Lemma 7.8 to do Exercise 7.22 without assuming that f is differentiable.

Chapter 8

Differentiable functions of several variables

First I have to admit that simplicity prevails: I will emphasize the story of convex *differentiable* functions of several variables. Therefore we need a very firm grip on the concept of differentiability for functions of several variables. Here the geometric intuition concerning the slope of the tangent line is not available. Following (7.12) one is forced to define the derivative of a map $f : \mathbb{R}^n \to \mathbb{R}^m$ at a point as an $m \times n$ matrix. Fortunately this approach, given in Definition 8.1, works out perfectly for the applications we have in mind, in particular a generalization of the one variable Newton-Raphson method (§8.1.1), local extrema for differentiable functions of several variables (§8.1.2) and the all important chain rule (Theorem 8.12).

As a prelude to the Karush-Kuhn-Tucker conditions, we introduce Lagrange multipliers for solving constrained optimization problems at the end of the chapter.

8.1 Differentiability

Recall that a function $f : (a, b) \to \mathbb{R}$ of one variable is differentiable at $x_0 \in (a, b)$ if there exists $c \in \mathbb{R}$ and $\delta > 0$ such that

$$f(x_0 + h) - f(x_0) = c\,h + \epsilon(h)h, \tag{8.1}$$

where $\epsilon : (-\delta, \delta) \to \mathbb{R}$ is a function continuous in 0 with $\epsilon(0) = 0$. The interpretation of c is the change in $f(x)$ measured to the change in x around x_0 for h small. The generalization of (8.1) to several variables is given below.

Definition 8.1. Let $f : U \to \mathbb{R}^m$ be a function with $U \subseteq \mathbb{R}^n$ an open subset. Then f is *differentiable* at $x_0 \in U$ if there exists an $m \times n$ matrix C, such that

$$f(x_0 + h) - f(x_0) = C\,h + \epsilon(h)\,|h|,$$

where $\epsilon : O \to \mathbb{R}^m$ is a function from an open subset $O \subseteq \mathbb{R}^n$ containing 0, such that ϵ is continuous in 0 with $\epsilon(0) = 0$. The function f is called differentiable if it is differentiable at every $x \in U$.

This definition is extremely compact and contains a wealth of information! To understand what is going on we need to recall the partial derivatives.

Definition 8.2. Let $f : U \to \mathbb{R}$ be a function, where U is an open subset of $\mathbb{R}^n$. Then we define the limit

$$\frac{\partial f}{\partial x_i}(v) := \lim_{\delta \to 0} \frac{f(v + \delta e_i) - f(v)}{\delta},$$

where $i = 1, \ldots, n$,

$$e_i = (0, \ldots, 0, \underset{\underset{i}{\uparrow}}{1}, 0, \ldots, 0)$$

is the *canonical i-th basis vector* in $\mathbb{R}^n$, $v \in U$ and $\delta \in \mathbb{R}$. This limit is called the *partial derivative* of f with respect to x_i at v.

Example 8.3. Consider the function $f : \mathbb{R}^2 \to \mathbb{R}$ given by

$$f(x_1, x_2) = x_1 x_2^2 + x_1.$$

Then

$$
\begin{aligned}
\frac{\partial f}{\partial x_2}(v) &= \lim_{\delta \to 0} \frac{f(x_1, x_2 + \delta) - f(x_1, x_2)}{\delta} \\
&= \lim_{\delta \to 0} \frac{x_1(x_2 + \delta)^2 + x_1 - (x_1 x_2^2 + x_1)}{\delta} \\
&= x_1 \lim_{\delta \to 0} \frac{(x_2 + \delta)^2 - x_2^2}{\delta} = x_1 \lim_{\delta \to 0} (2x_2 + \delta) = 2x_1 x_2,
\end{aligned}
$$

where $v = (x_1, x_2)$. This example illustrates that $\frac{\partial f}{\partial x_i}$ can be computed just like in the one variable case, when the other variables ($\neq x_i$) are treated as constants. Notice that

$$\frac{\partial}{\partial x_1} \frac{\partial f}{\partial x_2} = \frac{\partial}{\partial x_2} \frac{\partial f}{\partial x_1} = 2x_2.$$

A function $f : S \to \mathbb{R}^m$, where S is an arbitrary subset of $\mathbb{R}^n$, can be expressed as

$$f(x_1, \ldots, x_n) = \begin{pmatrix} f_1(x_1, \ldots, x_n) \\ \vdots \\ f_m(x_1, \ldots, x_n) \end{pmatrix}$$

where $f_i : S \to \mathbb{R}$ are (coordinate) functions from S to $\mathbb{R}$. We will implicitly use this convention in the following.

Proposition 8.4. *Let $f : U \to \mathbb{R}^m$ be a function with $U \subseteq \mathbb{R}^n$ an open subset. If f is differentiable at $x_0 \in U$, the partial derivatives*

$$\frac{\partial f_i}{\partial x_j}(x_0)$$

exist for $i = 1, \ldots, m$ and $j = 1, \ldots, n$ and the matrix C in Definition 8.1 is

$$C = \begin{pmatrix} \dfrac{\partial f_1}{\partial x_1}(x_0) & \cdots & \dfrac{\partial f_1}{\partial x_n}(x_0) \\ \vdots & \ddots & \vdots \\ \dfrac{\partial f_m}{\partial x_1}(x_0) & \cdots & \dfrac{\partial f_m}{\partial x_n}(x_0) \end{pmatrix}.$$

Proof. The j-th column in C is Ce_j. Putting $h = \delta e_j$ for $\delta \in \mathbb{R}$ in Definition 8.1 gives

$$f(x_0 + \delta e_j) - f(x_0) = \delta C e_j + \epsilon(\delta e_j)|\delta|.$$

The i-th coordinate of this identity of m-dimensional vectors can be written

$$f_i(x_0 + \delta e_j) - f_i(x_0) = \delta C_{ij} + \tilde{\epsilon}_i(\delta)\delta \tag{8.2}$$

where

$$\tilde{\epsilon}_i(\delta) = \begin{cases} \epsilon_i(\delta e_j)\dfrac{|\delta|}{\delta} & \text{if } \delta \neq 0 \\ 0 & \text{if } \delta = 0 \end{cases}$$

and (8.2) shows that $C_{ij} = \frac{\partial f_i}{\partial x_j}(x_0)$. $\qquad\square$

For a function $f : U \to \mathbb{R}$ with $U \subseteq \mathbb{R}^n$ an open subset, the partial derivative, if it exists for every $x \in U$, is a new function

$$\frac{\partial f}{\partial x_j} : U \to \mathbb{R}.$$

We will use the notation

$$\frac{\partial^2 f}{\partial x_i \partial x_j} := \frac{\partial}{\partial x_i} \frac{\partial f}{\partial x_j}$$

for the *iterated (second order) partial derivative.*

The first part of following result is a converse to Proposition 8.4. The second part contains the surprising symmetry of the second order partial derivatives under rather mild conditions. We will not go into the proof, which can be found in most calculus textbooks.

Theorem 8.5. *Let $f : U \to \mathbb{R}^m$ be a function with $U \subseteq \mathbb{R}^n$ an open subset. If the partial derivatives for f exist at every $x \in U$ with*

$$\frac{\partial f_i}{\partial x_j}$$

continuous (for $i = 1, \ldots, m$ and $j = 1, \ldots, n$), then f is differentiable. If the second order partial derivatives exist for a function $f : U \to \mathbb{R}$ and are continuous functions, then

$$\frac{\partial^2 f}{\partial x_i \partial x_j} = \frac{\partial^2 f}{\partial x_j \partial x_i}$$

for $i, j = 1, \ldots, n$.

You can check explicitly that the symmetry of the second order partial derivatives holds for functions $f(x, y) = \lambda x^\alpha y^\beta$, where $\lambda \in \mathbb{R}, \alpha, \beta \in \mathbb{N}$ and therefore also for sums of such functions (called polynomials in x and y). These functions are extremely well behaved unlike the horrific examples presented in Exercises 8.2 and 8.3.

The notation $f'(x_0)$ for the derivative at a point, x_0, for a function of one variable is very familiar. The derivative $f'(x_0)$ for a function $f : S \to \mathbb{R}^n$ at $x_0 \in S \subseteq \mathbb{R}^m$ is suitably generalized to a matrix (a 1×1-matrix is really nothing but a number).

Definition 8.6. Let $f : U \to \mathbb{R}^m$ be a function with $U \subseteq \mathbb{R}^n$ an open subset. If the partial derivatives

$$\frac{\partial f_i}{\partial x_j}(x)$$

exist for $x \in U$ (for $i = 1, \ldots, m$ and $j = 1, \ldots, n$), then we denote

$$f'(x) := \begin{pmatrix} \dfrac{\partial f_1}{\partial x_1}(x) & \cdots & \dfrac{\partial f_1}{\partial x_n}(x) \\ \vdots & \ddots & \vdots \\ \dfrac{\partial f_m}{\partial x_1}(x) & \cdots & \dfrac{\partial f_m}{\partial x_n}(x) \end{pmatrix}.$$

This matrix is called the *Jacobian matrix* or *the differential* for f at x. If $\frac{\partial f_i}{\partial x_j}$ are continuous functions on U, then f is called *continuously differentiable*.

8.1.1 The Newton-Raphson method for several variables

Let $f : \mathbb{R}^n \to \mathbb{R}^n$ be a differentiable function. Many applications lead to the problem of computing $\xi \in \mathbb{R}^n$ with $f(\xi) = 0$. Here a matrix variation of the Newton-Raphson method (see §7.5.1) may apply. The formulas are the same as in the one variable case. Definition 8.1 with $h = \xi - x_0$ shows that

$$f(\xi) - f(x_0) \quad \text{is close to} \quad f'(x_0)(\xi - x_0) \tag{8.3}$$

provided that $|h|$ is small. The approximate formula

$$-f(x_0) = f'(x_0)(\xi - x_0) \tag{8.4}$$

leading to the Newton-Raphson method is now an identity between vectors in $\mathbb{R}^n$ and $f'(x_0)$ is an $n \times n$ matrix. We only get a unique solution for ξ in (8.4) when $f'(x_0)$ is an invertible matrix. In the multivariable setting an iteration in the *Newton-Raphson method* then becomes

$$x_{n+1} = x_n - f'(x_n)^{-1} f(x_n) \tag{8.5}$$

provided that $f'(x_n)$ is invertible.

Example 8.7. For the differentiable map $f : \mathbb{R}^2 \to \mathbb{R}^2$ given by

$$f\left(\begin{pmatrix} x \\ y \end{pmatrix} \right) = \begin{pmatrix} y^2 - x^3 + x \\ y^3 - x^2 \end{pmatrix},$$

(8.5) is

$$\begin{pmatrix} x_{n+1} \\ y_{n+1} \end{pmatrix} = \begin{pmatrix} x_n \\ y_n \end{pmatrix} - \begin{pmatrix} -3x_n^2 + 1 & 2y_n \\ -2x_n & 3y_n^2 \end{pmatrix}^{-1} \begin{pmatrix} y_n^2 - x_n^3 + x_n \\ y_n^3 - x_n^2 \end{pmatrix}. \tag{8.6}$$

Starting out with $(x_0, y_0) = (1, 1)$, (8.6) approaches $(1.46107, 1.2879)$ in a few steps. This turns out to be a good approximation to a zero for the equations

$$y^2 - x^3 + x = 0$$
$$y^3 - x^2 = 0.$$

For $(x_0, y_0) = (-1, 1)$, (8.6) approaches $(-0.471074, 0.605423)$, which is another good approximation to a zero. If (8.6) is started with $(x_0, y_0) = (-1, -1)$, we get $(x_1, y_1) = (-1.5, 0)$. Here the matrix

$$\begin{pmatrix} -3x_1^2 + 1 & 2y_1 \\ -2x_1 & 3y_1^2 \end{pmatrix}$$

is singular and the process comes to an unsuccessful halt.

The Newton-Raphson method for several variables will be applied later (see §10.5) in giving an algorithm for solving convex optimization problems.

8.1.2   *Local extrema for functions of several variables*

We have deviated slightly from the usual notation. For a function $f : U \to \mathbb{R}$ as in Definition 8.6, the differential $f'(x_0)$ at $x_0 \in U$ is called *the gradient* for f at x_0. It is usually denoted $\nabla f(x_0)$. Now we give the multivariable generalization of Definition 7.10.

Definition 8.8. Let $f : U \to \mathbb{R}$ be a function, where $U \subseteq \mathbb{R}^n$ is an open subset. Suppose that the partial derivatives exist at $x_0 \in U$. Then x_0 is called a *critical point* for f if $\nabla f(x_0) = 0$.

Recall Definition 7.5 of a local extremum for a function $f : S \to \mathbb{R}$, where S is any subset of $\mathbb{R}^n$.

If x_0 is not a critical point for f we can use the gradient vector to move in a direction making f strictly smaller/larger. This is very important for optimization problems. Here is the natural generalization of Lemma 7.11.

Lemma 8.9. *Let $f : U \to \mathbb{R}$ be a differentiable function, where $U \subseteq \mathbb{R}^n$ is an open subset. Suppose that $u \in \mathbb{R}^n$ and $\nabla f(x_0)u < 0$ for $x_0 \in U$. Then*

$$f(x_0 + \lambda u) < f(x_0)$$

for $\lambda > 0$ small.

Proof. By the differentiability of f,

$$f(x_0 + u) - f(x_0) = \nabla f(x_0)u + \epsilon(u)|u|,$$

where $\epsilon : \mathbb{R}^n \to \mathbb{R}$ is a function satisfying $\epsilon(h) \to 0$ for $h \to 0$. For $\lambda > 0$ with $x_0 + \lambda u \in U$ we have

$$f(x_0 + \lambda u) - f(x_0) = \lambda(\nabla f(x_0)u + \epsilon(\lambda u)|u|).$$

When λ tends to zero from the right, it follows that $f(x_0 + \lambda u) - f(x_0) < 0$ for small $\lambda > 0$. $\qquad\square$

Lemma 8.9 leads to a stepwise method for approximating a local minimum for the function $f : U \to \mathbb{R}$. If $\nabla f(x_0) \neq 0$, a natural choice of u is $-\nabla f(x_0)$. Here we get the existence of $\lambda_0 > 0$, such that $f(x_1) < f(x_0)$ with $x_1 := x_0 + \lambda_0 u$. The same step may be repeated with x_1 and $u = -\nabla f(x_1)$ provided that $\nabla f(x_1) \neq 0$. This iterative procedure is called the method of *steepest descent* and goes back at least to 1847, where it appeared in a paper by Cauchy on the solutions of systems of equations.

Notice that Lemma 8.9 is silent on the size of $\lambda > 0$. It merely states that $f(x_0 + \lambda u) < f(x_0)$ for some $\lambda > 0$. To analyze the size of λ one may consider the one variable function

$$g(t) = f(x_0 + tu)$$

and look for the first positive solution λ_0 (if any) to $g'(t) = 0$.

Corollary 8.10. *Let $f : U \to \mathbb{R}$ be a differentiable function, where $U \subseteq \mathbb{R}^n$ is an open subset. If $x_0 \in U$ is a local extremum, then x_0 is a critical point for f.*

Proof. Suppose that $\nabla f(x_0) \neq 0$. If x_0 is a local minimum, then we may use $u = -\nabla f(x_0)$ in Lemma 8.9 to deduce that $f(x_0 + \lambda u) < f(x_0)$ for $\lambda > 0$ small. This contradicts the local minimality of x_0. If x_0 is a local maximum we can apply Lemma 8.9 with $-f$ and $u = \nabla f(x_0)$ to reach a similar contradiction. Therefore $\nabla f(x_0) = 0$ and x_0 is a critical point for f. $\qquad\square$

Example 8.11. Consider the function $f : \mathbb{R}^2 \to \mathbb{R}$ given by

$$f(x, y) = x^4 y^2 + x^2 y^4 - 3x^2 y^2.$$

Here

$$\nabla f = \left(2xy^2(2x^2 + y^2 - 3), \ 2x^2 y(x^2 + 2y^2 - 3)\right)$$

and (x_0, y_0) is a critical point for f if and only if $x_0 = 0$, $y_0 = 0$ or $x_0 = y_0 = \pm 1$. In fact (see Exercise 7.3),

$$\min\{f(x, y) \,|\, (x, y) \in \mathbb{R}^2\} = -1$$

corresponding to the four global minima $(\pm 1, \pm 1)$.

8.2   The chain rule

Recall the chain rule for functions of one variable. Here we have functions $f : (a, b) \to \mathbb{R}$ and $g : (c, d) \to \mathbb{R}$, such that $g(x) \in (a, b)$ for $x \in (c, d)$. If g is differentiable at $x_0 \in (c, d)$ and f is differentiable at $g(x_0) \in (a, b)$, the chain rule says that $f \circ g$ is differentiable at x_0 with

$$(f \circ g)'(x_0) = f'(g(x_0))g'(x_0).$$

This rule generalizes verbatim to functions of several variables:

$$(f \circ g)'(x_0) = f'(g(x_0))g'(x_0)$$

for compatible multivariate functions f and g when you replace usual multiplication by matrix multiplication. Here is the precise statement with proof.

Theorem 8.12. *Let $f : U \to \mathbb{R}^m$ and $g : V \to \mathbb{R}^n$ with $U \subseteq \mathbb{R}^n$, $V \subseteq \mathbb{R}^l$ open subsets and $g(V) \subseteq U$. If g is differentiable at $x_0 \in V$ and f is differentiable at $g(x_0) \in U$, then $f \circ g$ is differentiable at x_0 with*

$$(f \circ g)'(x_0) = f'(g(x_0))g'(x_0).$$

Proof.   This proof is formally tedious but not conceptually difficult — it consists in working diligently with the ϵ-functions in the definition of differentiability. We will apply Definition 8.1 and its notation. Differentiability of g at x_0 means that

$$g(x_0 + h) - g(x_0) = g'(x_0)h + \epsilon_g(h)|h|,$$

for $h \in O_g$, where O_g is the open subset containing $0 \in \mathbb{R}^l$ in the definition of $\epsilon_g : O_g \to \mathbb{R}^n$. Differentiability of f at $g(x_0)$ means

$$f(g(x_0) + k) - f(g(x_0)) = f'(g(x_0))k + \epsilon_f(k)|k|, \tag{8.7}$$

for $k \in O_f$, where O_f is the open subset containing $0 \in \mathbb{R}^n$ in the definition of $\epsilon_f : O_f \to \mathbb{R}^l$. Therefore

$$
\begin{aligned}
f(g(x_0 + h)) &- f(g(x_0)) \\
&= f(g(x_0) + g'(x_0)h + \epsilon_g(h)|h|) - f(g(x_0)) \qquad (8.8) \\
&= f'(g(x_0))g'(x_0)h + \epsilon_1(h)|h|,
\end{aligned}
$$

where

$$
\epsilon_1(h) = \epsilon_f(g'(x_0)h + \epsilon_g(h)|h|) \left| g'(x_0)\frac{h}{|h|} + \epsilon_g(h) \right|
$$

provided that $g'(x_0)h + \epsilon_g(h)|h| \in O_f$. This holds for some open subset $O'_g \subseteq O_g$. One can check that $\epsilon_1 : O'_g \to \mathbb{R}^m$ satisfies $\epsilon_1(h) \to 0$ for $h \to 0$ (even though ϵ_1 is not defined in 0, we can put $\epsilon_1(0) = 0$ to comply with Definition 8.1). This shows that $f \circ g$ is differentiable at x_0 and that the coefficient matrix, $f'(g(x_0))g'(x_0)$ in front of h in (8.8) must equal $(f \circ g)'(x_0)$. $\qquad\square$

The chain rule is extremely important and useful. The compact notation $f'(g(x_0))g'(x_0)$ is a useful mnemonic: think of the one variable formula in terms of matrices of the right dimensions.

Example 8.13. Let $g : \mathbb{R}^2 \to \mathbb{R}$ and $\varphi : \mathbb{R} \to \mathbb{R}$ be two differentiable functions. Then

$$
f(t) = g(t, \varphi(t))
$$

is a differentiable function $\mathbb{R} \to \mathbb{R}$ by Theorem 8.12. The chain rule also gives a formula for computing $f'(t)$. First notice that f may be viewed as $g \circ \varphi_1$, where $\varphi_1(t) = (t, \varphi(t))$. Therefore

$$
\varphi_1'(t) = \begin{pmatrix} 1 \\ \varphi'(t) \end{pmatrix} \quad \text{and} \quad g'(v_0) = \left(\frac{\partial g}{\partial x}(v_0), \frac{\partial g}{\partial y}(v_0) \right)
$$

for $v_0 \in \mathbb{R}^2$ and $t \in \mathbb{R}$ giving the formula

$$
\begin{aligned}
f'(t) = g'(\varphi_1(t))\varphi_1'(t) &= \left(\frac{\partial g}{\partial x}(\varphi_1(t)), \frac{\partial g}{\partial y}(\varphi_1(t)) \right) \begin{pmatrix} 1 \\ \varphi'(t) \end{pmatrix} \\
&= \frac{\partial g}{\partial x}(t, \varphi(t)) + \varphi'(t) \frac{\partial g}{\partial y}(t, \varphi(t)).
\end{aligned}
$$

8.3   Lagrange multipliers

What is the minimal value of the (linear) function $f(x, y) = x + y$ subject to $g(x, y) = x^2 + y^2 - 1 = 0$? This is an example of a minimization problem with a non-linear constraint function.

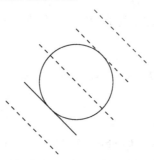

Figure 8.1: Level curves $\{(x, y) \in \mathbb{R}^2 \mid x + y = c\}$ for five values of c compared to the constraint set $\{(x, y) \in \mathbb{R}^2 \mid x^2 + y^2 - 1 = 0\}$ in the minimization problem $\min\{x + y \mid x^2 + y^2 - 1 = 0\}$.

Let us have a closer look at the optimization problem

$$\min\{f(x, y) \mid g(x, y) = 0\} \tag{8.9}$$

for general differentiable functions f and g. We are studying the values of f along the set S of points $(x, y) \in \mathbb{R}^2$ satisfying $g(x, y) = 0$. A local minimum in (8.9) is by Definition 7.5 a point $v = (x_0, y_0) \in S$, such that

$$f(v) \leq f(u) \tag{8.10}$$

for every $u \in U \cap S$, where $U \subseteq \mathbb{R}^2$ is an open subset containing v. Using the same idea as in Example 1.2, (8.9) can be reformulated as the task of finding the minimal value $c = c_0 \in \mathbb{R}$, such that the two level curves

$$
\begin{aligned}
f(x, y) &= c \\
g(x, y) &= 0
\end{aligned}
\tag{8.11}
$$

intersect i.e., there exists (x_0, y_0), such that $f(x_0, y_0) = c_0$ and $g(x_0, y_0) = 0$ and (8.11) has no solution for $c < c_0$. Intuitively this means that the level curves $f(x, y) = c$ and $g(x, y) = 0$ barely touch each other at (x_0, y_0). Geometrically, they should have a common tangent i.e., there should exist $\lambda \in \mathbb{R}$ so that

$$
\begin{aligned}
g(x_0, y_0) &= 0 \\
\nabla f(x_0, y_0) &= \lambda \nabla g(x_0, y_0).
\end{aligned}
\tag{8.12}
$$

This situation is depicted by the non-dashed level curve in Figure 8.1. Another way of stating (8.12), is that $(x_0, y_0, \lambda) \in \mathbb{R}^3$ should be a critical point for the (Lagrangian) function

$$L(x, y, \lambda) = f(x, y) + \lambda g(x, y)$$

defined on $\mathbb{R}^3$. In the concrete case $f(x, y) = x + y$ and $g(x, y) = x^2 + y^2 - 1$, we end up with the following three equations

$$x^2 + y^2 = 1$$
$$2x\lambda = 1$$
$$2y\lambda = 1$$

in the unknowns x, y, λ. This is an example of a system of three non-linear equations in three unknowns.

We will show why these intuitive remarks really work under certain assumptions by using the so-called *implicit function theorem* from analysis as a black box.

The two variable case

In one variable calculus a local extremum x_0 for a differentiable function f satisfies $f'(x_0) = 0$. This fact is used extensively when computing local extrema in the one variable case. We are looking for a similar characterization of a local extremum in (8.9). Here the trick is to consider the one variable function $f(t, x(t))$, where $x : (a, b) \to \mathbb{R}$ is a carefully selected differentiable function satisfying $g(t, x(t)) = 0$.

For a given point $v = (t_0, y_0)$ with $g(v) = 0$ we may ask if there exists a differentiable function $x : (t_0 - \epsilon, t_0 + \epsilon) \to \mathbb{R}$ with $x(t_0) = y_0$ such that $g(t, x(t)) = 0$.

Example 8.14. Consider as an example the function $g(x, y) = x^2 + y^2 - 1$ and the point $v = (0, 1)$ with $g(v) = 0$. Here the differentiable function $x : (-1, 1) \to \mathbb{R}$ given by

$$x(t) = \sqrt{1 - t^2}$$

satisfies $x(0) = 1$ and $g(t, x(t)) = 0$.

On the other hand, for the point $v = (1, 0)$ it seems not so obvious to find a differentiable function $x : (1 - \epsilon, 1 + \epsilon) \to \mathbb{R}$ for $\epsilon > 0$ with $x(1) = 0$ and $g(t, x(t)) = 0$. This is related to the fact that

$$\frac{\partial g}{\partial y}(v) = 0.$$

Theorem 8.15. *If* $g : \mathbb{R}^2 \to \mathbb{R}$ *is a continuously differentiable function,* $v = (t_0, y_0) \in \mathbb{R}^2$ *is a point with* $g(v) = 0$ *and*

$$\frac{\partial g}{\partial y}(v) \neq 0,$$

there exists a differentiable function $x : (t_0 - \epsilon, t_0 + \epsilon) \to \mathbb{R}$ *for* $\epsilon > 0$ *such that* $x(t_0) = y_0$ *and* $g(t, x(t)) = 0$.

Proof.   The result follows from the implicit function theorem. We will not go into the details here.   □

Now suppose in (8.9) that $v = (t_0, y_0)$ is a local minimum with

$$\frac{\partial g}{\partial y}(v) \neq 0.$$

Then there exists a differentiable function $x : (t_0 - \epsilon, t_0 + \epsilon) \to \mathbb{R}$ for $\epsilon > 0$ with $x(t_0) = y_0$ and $g(t, x(t)) = 0$ by Theorem 8.15. By definition of a local minimum for (8.9), it follows that t_0 must be a local minimum for the one variable differentiable function $F : (t_0 - \epsilon, t_0 + \epsilon) \to \mathbb{R}$ given by

$$F(t) = f(t, x(t)).$$

Therefore $F'(t_0) = 0$ by Lemma 7.11. But here Example 8.13 shows that

$$F'(t_0) = \frac{\partial f}{\partial x}(v) + \frac{\partial f}{\partial y}(v)x'(t_0) = 0. \tag{8.13}$$

Since $g(t, x(t)) = 0$, the chain rule also applies to the function $G(t) = g(t, x(t))$ and

$$G'(t_0) = \frac{\partial g}{\partial x}(v) + \frac{\partial g}{\partial y}(v)x'(t_0) = 0. \tag{8.14}$$

If you compare (8.13) with (8.14) you will see that we have found a non-zero vector $w \in \mathbb{R}^2$, such that

$$\nabla f(v)w = 0 \quad \text{and} \quad \nabla g(v)w = 0.$$

Therefore $\nabla f(v)$ and $\nabla g(v)$ must be linearly dependent i.e., there exists $\lambda \in \mathbb{R}$, such that

$$\nabla f(v) = \lambda \nabla g(v). \tag{8.15}$$

Here λ is referred to as a *Lagrange multiplier* at v.

Let us finish this section by solving the non-linear equations coming from the concrete example mentioned earlier.

Example 8.16. If $f(x, y) = x + y$ and $g(x, y) = x^2 + y^2 - 1$, then

$$\nabla f = (1, 1) \quad \text{and} \quad \nabla g = (2x, 2y)$$

and (8.15) becomes

$$1 - 2x\lambda = 0$$
$$1 - 2y\lambda = 0.$$

When we add the constraint $g(x, y) = 0$, we get three equations with the three unknowns x, y and λ:

$$1 - 2x\lambda = 0$$
$$1 - 2y\lambda = 0$$
$$x^2 + y^2 - 1 = 0.$$

The first two equations give $x = y = 1/(2\lambda)$, which is substituted into the third giving

$$\lambda = \pm \frac{1}{\sqrt{2}}.$$

For these two values of λ we have the following candidates for local optima:

$$\pm \begin{pmatrix} \sqrt{2}/2 \\ \sqrt{2}/2 \end{pmatrix}.$$

You can see that the $+$ $(-)$ sign corresponds to the global maximum (minimum).

The general case and the Lagrangian

Let

$$f, g_1, \ldots, g_m : \mathbb{R}^n \to \mathbb{R}$$

be differentiable functions of n variables with $n > m$. Consider the optimization problem

$$\min\{f(x) \,|\, x \in S\}, \tag{8.16}$$

where

$$S = \left\{ (x_1, \ldots, x_n) \in \mathbb{R}^n \; \middle| \; \begin{array}{c} g_1(x_1, \ldots, x_n) = 0 \\ \vdots \\ g_m(x_1, \ldots, x_n) = 0 \end{array} \right\}.$$

With an additional assumption, the optimal points for (8.16) can be found among the critical points of the differentiable function $L : \mathbb{R}^{n+m} \to \mathbb{R}$ defined by

$$L(x_1, \ldots, x_n, \lambda_1, \ldots, \lambda_m)$$
$$:= f(x_1, \ldots, x_n) + \lambda_1 g_1(x_1, \ldots, x_n) + \cdots + \lambda_m g_m(x_1, \ldots, x_n).$$

This function is called the *Lagrangian function* or simply the *Lagrangian* associated to (8.16). The purpose of the Lagrangian is to transform the constrained optimization problem (8.16) into an unconstrained one, where Corollary 8.10 applies. The Lagrangian is named after the French mathematician Lagrange.

Figure 8.2: Joseph Louis Lagrange (1736–1813). French mathematician.

If $(x_0, \lambda_1, \ldots, \lambda_m)$ is a critical point for the Langrangian with $x_0 \in S$, then $\lambda_1, \ldots, \lambda_m$ are called *Lagrange multipliers* at x_0.

Theorem 8.17. *Suppose that $x_0 \in S$ is a local extremum for (8.16) and that*

$$\nabla g_1(x_0), \ldots, \nabla g_m(x_0)$$

are linearly independent. Then there exists $\lambda_1, \ldots, \lambda_m \in \mathbb{R}$, such that $(x_0, \lambda_1, \ldots, \lambda_m) \in \mathbb{R}^{n+m}$ is a critical point for the Lagrangian of (8.16).

Proof. The critical point condition for the Lagrangian translates into the existence of $\lambda_1, \ldots, \lambda_m$ such that

$$\nabla f(x_0) = \lambda_1 \nabla g_1(x_0) + \cdots + \lambda_m \nabla g_m(x_0).$$

Inspired by the exposition in [Iversen (1988)], we sketch the details in the proof. The deeper underlying result is the existence of a differentiable map

$$\theta : U \to \mathbb{R}^n,$$

where $U \subseteq \mathbb{R}^{n-m}$ is an open subset, such that

(1) There exists an open subset $V \subseteq \mathbb{R}^n$ with $x_0 \in V$ and

$$\theta(U) = S \cap V.$$

(2) θ is injective.
(3) $\theta'(x)$ has full rank $n - m$ for every $x \in U$.

This is a consequence of the linear independence assumption on $\nabla g_1(x_0), \ldots, \nabla g_m(x_0)$ and can be proved from the implicit function theorem in analysis. Starting with these assumptions, the proof is formally similar to the two variable case with some more linear algebra added: let $F : U \to \mathbb{R}$ be the differentiable function given by

$$F = f \circ \theta. \tag{8.17}$$

If $\theta(a) = x_0$, then a is a local extremum for F and $\nabla F(a) = 0$ by Theorem 8.10. Applying the chain rule to (8.17) we get

$$\nabla F(a) = \nabla f(x_0)\theta'(a) = 0.$$

Similarly, the chain rule applied to

$$g_i \circ \theta = 0,$$

gives

$$\nabla g_i(x_0)\theta'(a) = 0$$

for $i = 1, \ldots, m$. From here the rest is linear algebra: the subspace

$$K = \{u \in \mathbb{R}^n \mid u^t\theta'(a) = 0\}$$

of $\mathbb{R}^n$ has dimension m, since the $n \times (n - m)$ matrix $\theta'(a)$ has rank $n - m$ by assumption. But $\nabla g_1(x_0), \ldots, \nabla g_m(x_0) \in K$ must form a basis of K, as they are linearly independent. Since $\nabla f(x_0) \in K$ we must have

$$\nabla f(x_0) = \lambda_1 \nabla g_1(x_0) + \cdots + \lambda_m \nabla g_m(x_0)$$

for suitable $\lambda_1, \ldots, \lambda_m \in \mathbb{R}$. $\qquad\square$

Remark 8.18. Using Theorem 8.17 in solving (8.16) one needs to solve $m + n$ equations:

$$g_1(x_1, \ldots, x_n) = 0$$
$$\vdots$$
$$g_m(x_1, \ldots, x_n) = 0$$
$$\frac{\partial f}{\partial x_1}(x_1, \ldots, x_n) = \lambda_1 \frac{\partial g_1}{\partial x_1}(x_1, \ldots, x_n) + \cdots + \lambda_m \frac{\partial g_m}{\partial x_1}(x_1, \ldots, x_n)$$
$$\vdots$$
$$\frac{\partial f}{\partial x_n}(x_1, \ldots, x_n) = \lambda_1 \frac{\partial g_1}{\partial x_n}(x_1, \ldots, x_n) + \cdots + \lambda_m \frac{\partial g_m}{\partial x_n}(x_1, \ldots, x_n)$$

in the $m + n$ unknowns $x_1, \ldots, x_n, \lambda_1, \ldots, \lambda_m$. This can be quite hard, if not impossible.

Example 8.19. Find the maximum and minimum values of $f(x, y, z) = x - 2y + z$, when $x^2 + y^2 + z^2 = 1$ and $x + y + z = 0$. Here the equations in Remark 8.18 becomes

$$x^2 + y^2 + z^2 - 1 = 0$$
$$x + y + z = 0$$
$$1 = 2\lambda x + \mu \qquad\qquad (8.18)$$
$$-2 = 2\lambda y + \mu$$
$$2 = 2\lambda z + \mu.$$

The linear independence of the gradients in Theorem 8.17 is satisfied, since

$$\begin{pmatrix} 2x \\ 2y \\ 2z \end{pmatrix} \quad \text{and} \quad \begin{pmatrix} 1 \\ 1 \\ 1 \end{pmatrix}$$

are linearly dependent if and only if $x = y = z$. This is impossible under the constraints $x^2 + y^2 + z^2 = 1$ and $x + y + z = 0$. Eliminating z you can deduce that $\mu = 1/3$ from (8.18). Expressing x, y and z in terms of λ and inserting in $x^2 + y^2 + z^2 = 1$ gives

$$\lambda = \pm\sqrt{\frac{13}{6}}.$$

For $\lambda = \sqrt{13/6}$ we get

$$\begin{pmatrix} x \\ y \\ z \end{pmatrix} = \begin{pmatrix} \sqrt{2/39} \\ -\sqrt{7/78} \\ \sqrt{5/78} \end{pmatrix}$$

which turns out to be the maximum of $f(x, y, z)$ under the constraints. You should compute the minimum and fill in the missing details in the computations above (Exercise 8.5). This example shows that the (non-linear!) equations in Remark 8.18 can be approachable, but that some tricks may be involved in solving them.

8.4 The arithmetic-geometric inequality revisited

The method of Lagrange multipliers as presented in Theorem 8.17 is quite powerful. As an example, we use it to give a new proof of the inequality

$$\sqrt[n]{x_1 \cdots x_n} \leq \frac{x_1 + \cdots + x_n}{n}$$

between the geometric and arithmetic means for $x_1, \ldots, x_n \geq 0$. Suppose that

$$\frac{x_1 + \cdots + x_n}{n} = a \geq 0. \tag{8.19}$$

We will approach the proof by investigating the extrema of $x_1 \cdots x_n$ under the assumption (8.19). Let $g(x_1, \ldots, x_n) = x_1 + \cdots + x_n - na$ for fixed $a > 0$. Then Theorem 8.17 applies to

$$\max_{x \in S} x_1 \cdots x_n, \tag{8.20}$$

where $S = \{(x_1, \ldots, x_n) \mid g(x_1, \ldots, x_n) = 0\}$. The equations in Remark 8.18 become

$$x_1 + \cdots + x_n = na$$
$$x_2 x_3 \cdots x_n = \lambda$$
$$x_1 x_3 \cdots x_n = \lambda$$
$$\vdots$$
$$x_1 x_2 \cdots x_{n-1} = \lambda.$$

Therefore

$$x_1 = x_2 = \cdots = x_n$$

and $x_1 = x_2 = \cdots = x_n = a$, since we may assume that $x_1, \ldots, x_n > 0$. So if $(x_1, \ldots, x_n)$ is a local maximum in (8.20) for $a > 0$, the inequality follows. What if $a = 0$ or some $x_i = 0$ for $i = 1, \ldots, n$?

Later on, we will encounter Theorem 8.17 in a beautiful proof due to Cauchy of the fact that every real symmetric matrix has a real eigenvalue (see Theorem 9.24).

8.5   Exercises

Exercise 8.1. Consider $f : \mathbb{R} \to \mathbb{R}^3$ and $g : \mathbb{R}^3 \to \mathbb{R}$ given by

$$f(t) = \begin{pmatrix} t \\ t^2 \\ t^3 \end{pmatrix} \quad \text{and} \quad g\begin{pmatrix} x \\ y \\ z \end{pmatrix} = x^2 + 3y^6 + 2z^5.$$

Compute $(g \circ f)'(t)$ using the chain rule and check the result with an explicit computation of the derivative of $g \circ f : \mathbb{R} \to \mathbb{R}$.

Exercise 8.2. Consider the function $f : \mathbb{R} \to \mathbb{R}$ defined by

$$f(x) = \begin{cases} x^2 \sin(\frac{1}{x}), & \text{if } x \neq 0 \\ 0, & \text{if } x = 0. \end{cases}$$

Prove that f is differentiable. Is f' a continuous function?

Exercise 8.3. Let $f : \mathbb{R}^2 \to \mathbb{R}$ be given by

$$f(x, y) = \begin{cases} \dfrac{xy(x^2 - y^2)}{x^2 + y^2} & \text{if } (x, y) \neq (0, 0) \\ 0 & \text{if } (x, y) = (0, 0). \end{cases}$$

Verify that

$$\frac{\partial f}{\partial x}(v) = \frac{y\left(x^4 + 4x^2y^2 - y^4\right)}{\left(x^2 + y^2\right)^2}$$

$$\frac{\partial f}{\partial y}(v) = \frac{x\left(x^4 - 4x^2y^2 - y^4\right)}{\left(x^2 + y^2\right)^2}$$

and show that

$$\left|\frac{\partial f}{\partial x}(v)\right| \leq 2|y|$$

$$\left|\frac{\partial f}{\partial y}(v)\right| \leq 2|x|$$

for $v = (x, y) \neq (0, 0)$. Show that

$$\frac{\partial f}{\partial x}(0, y) = \begin{cases} -y & \text{if } y \neq 0 \\ 0 & \text{if } y = 0, \end{cases}$$

$$\frac{\partial f}{\partial y}(x, 0) = \begin{cases} x & \text{if } x \neq 0 \\ 0 & \text{if } x = 0. \end{cases}$$

Prove that f is a differentiable function using Theorem 8.5. Verify that

$$\frac{\partial^2 f}{\partial x \partial y}(0,0) = 1 \quad \text{and} \quad \frac{\partial^2 f}{\partial y \partial x}(0,0) = -1.$$

What does this have to do with the second part of Theorem 8.5?

Exercise 8.4. How would you find a numerical approximation to a root of the equation $x = \cos(x)$? Here is a suggestion for a strategy you may not have seen before. First introduce an extra parameter t parametrizing the family

$$x = t \cos(x)$$

of equations. Of course you know the solution for $t = 0$. You want it for $t = 1$. You can do this by following a function $x : [0,1] \to \mathbb{R}$ with $x(0) = 0$ satisfying

$$x(t) = t \cos(x(t)).$$

Show that such a function satisfies

$$x'(t) = \frac{\cos(x(t))}{1 + t \sin(x(t))}.$$

Use the approximation

$$x(t_0 + \delta) = x(t_0) + \delta x'(t_0)$$

with step size $\delta = 0.1$ in this method to find an approximate solution to $x = \cos(x)$. Do you see a way of improving this numerical method (apart from choosing a smaller δ)?

Note. The type of algorithm hinted to in Exercise 8.4 is called a *homotopy method*. The terminology comes from the parameter t which moves from $t = 0$ (the solution you know) to $t = 1$ (the solution you want).

Exercise 8.5. Fill in the details in the computations of Example 8.19.

Exercise 8.6. Maximize $x^2 + y^2$ subject to $x^2 + xy + y^2 = 4$. Does this problem have a geometric interpretation?

Exercise 8.7. Fill in the details in the proof of the inequality between the geometric and arithmetic means given in §8.4.

Exercise 8.8. A rectangular box has side lengths x, y and z. What is its maximal volume when we assume that (x, y, z) lies on the plane

$$\frac{x}{a} + \frac{y}{b} + \frac{z}{c} = 1$$

for $a, b, c > 0$.

Exercise 8.9. A company is planning to produce a box with volume $2\,\mathrm{m}^3$. For design reasons it needs different materials for the sides, top and bottom. The cost of the materials per square meter is 1 dollar for the sides, 1.5 dollars for the bottom and the top. Find the measurements of the box minimizing the production costs.

Exercise 8.10. What is wrong with the following proof of Theorem 8.17?

Proof.   Suppose there exists a non-zero vector $v \in \mathbb{R}^n$ with $\nabla f(v_0)v \neq 0$ and

$$\nabla g_1(v_0)v = \cdots = \nabla g_m(v_0)v = 0.$$

Now define one variable differentiable functions by $F(t) = f(v_0 + tv)$ and $G_i(t) = g_i(v_0 + tv)$ for $i = 1, \ldots, m$. By the chain rule one gets

$$F'(t) = \nabla f(v_0)v \neq 0$$
$$G_i'(t) = \nabla g_i(v_0)v = 0.$$

This implies that $g_i(v_0 + tv) = g_i(v_0) = 0$ for every $t \in \mathbb{R}$. Therefore v_0 cannot be an extremum under these conditions and it follows that $\nabla f(v_0)$ lies in the linear span of $\nabla g_1(v_0), \ldots, \nabla g_m(v_0)$. $\square$

Chapter 9

Convex functions of several variables

In this chapter we will enter into convex functions of several variables i.e., convex functions $f : C \to \mathbb{R}$, where C is a convex subset of $\mathbb{R}^n$ for $n \geq 1$. A lot of the results here make use of the theory for $n = 1$. A notable exception is the existence of subgradients (Theorem 9.3) for a convex function. To prepare for this encounter, you may want to begin looking for "tangents" for the absolute value function $f(x) = |x|$ at $x = 0$ or more ambitiously solve Exercise 9.7 from scratch only using the definition of a convex function.

The focus is, however, primarily on nice differentiable convex functions of several variables. Here Theorem 9.5 is the natural generalization of the fact that $f : (a, b) \to \mathbb{R}$ is convex if and only if $f''(x) \geq 0$. For $f : U \to \mathbb{R}$ with $U \subseteq \mathbb{R}^n$ an open subset, the generalization of $f''(x)$ is a symmetric $n \times n$ matrix denoted $\nabla^2 f(x)$ (called the Hessian of f at x) and ≥ 0 is replaced by "positive semidefinite".

It is my experience teaching convex functions to second year undergraduates, that positive semidefinite matrices and quadratic forms rarely are fully understood or appreciated from a first linear algebra course. A substantial part of this chapter is an attempt to remedy this gap. If you are comfortable with this important part of linear algebra or know the classical reduction of a symmetric matrix (quadratic form) to a diagonal matrix (sum of squares of linear forms), you are encouraged to skip the relevant sections.

9.1   Subgradients

We begin be generalizing Theorem 7.21 in order to motivate the definition of the subgradient at a point for a general convex function.

Theorem 9.1. *Let $f : U \to \mathbb{R}$ be a differentiable function, where $U \subseteq \mathbb{R}^n$ is an open convex subset. Then f is convex if and only if*

$$f(x) \geq f(x_0) + \nabla f(x_0)(x - x_0) \tag{9.1}$$

for every $x, x_0 \in U$.

Proof.   Suppose that (9.1) holds and let $x_t = (1-t)x_0 + tx$ with $0 \leq t \leq 1$, where $x_0, x \in U$. To prove that f is convex we must verify the inequality

$$f(x_t) \leq (1 - t)f(x_0) + tf(x). \tag{9.2}$$

Let $\xi = \nabla f(x_t)$. Then

$$f(x) \geq f(x_t) + \xi(1 - t)(x - x_0)$$
$$f(x_0) \geq f(x_t) - \xi t(x - x_0)$$

by (9.1). If you multiply the first inequality by t, the second by $1 - t$ and then add the two, you get (9.2). Suppose on the other hand that f is a convex function. Let $x_0, x \in U$. Since U is an open subset, it follows that $(1 - t)x_0 + tx \in U$ for $t \in I = (-\delta, 1 + \delta)$, where $\delta > 0$ is sufficiently small. Now define the function $g : I \to \mathbb{R}$ by

$$g(t) = f((1 - t)x_0 + tx) = f(x_0 + t(x - x_0)).$$

Being the composition of two differentiable functions, g is differentiable. Suppose that $0 \leq \alpha \leq 1$ and $t_1, t_2 \in I$. Then

$$\begin{aligned}
g((1 - \alpha)t_1 + \alpha t_2) &= f(x_0 + ((1 - \alpha)t_1 + \alpha t_2)(x - x_0)) \\
&= f((1 - \alpha)(x_0 + t_1(x - x_0)) + \alpha(x_0 + t_2(x - x_0))) \\
&\leq (1 - \alpha)f(x_0 + t_1(x - x_0)) + \alpha f(x_0 + t_2(x - x_0)) \\
&= (1 - \alpha)g(t_1) + \alpha g(t_2)
\end{aligned}$$

showing that g is a convex function. By Theorem 7.21,

$$g(1) \geq g(0) + g'(0),$$

which translates into

$$f(x) \geq f(x_0) + \nabla f(x_0)(x - x_0)$$

by using the chain rule in computing $g'(0)$. $\square$

It is worthwhile to record a somewhat stunning consequence of Theorem 9.1.

Corollary 9.2. *Let* $f : U \to \mathbb{R}$ *be a differentiable convex function, where* U *is an open convex subset of* $\mathbb{R}^n$. *A point* $x_0 \in U$ *is a global minimum for* f *if and only if* $\nabla f(x_0) = 0$.

Proof.   If $x_0 \in U$ is a global minimum, then $\nabla f(x_0) = 0$ by Theorem 8.10. If on the other hand $\nabla f(x_0) = 0$, then x_0 is a global minimum by Theorem 9.1. $\square$

Now we are ready to introduce the concept of a subgradient at a point of an arbitrary convex function.

Let S be a subset of $\mathbb{R}^n$. A vector $\xi \in \mathbb{R}^n$ is called a *subgradient* for the function $f : S \to \mathbb{R}$ at $x_0 \in S$ if

$$f(x) \geq f(x_0) + \xi^t(x - x_0)$$

for every $x \in S$. The set of all subgradients of f at x_0 is called the *subdifferential* for f at x_0. It is denoted $\partial f(x_0)$. Theorem 9.1 shows that

$$\nabla f(x_0) \in \partial f(x_0)$$

if f is differentiable. Using the definition of differentiability at a point x_0 it is not too difficult to prove that a convex and differentiable function $f : U \to \mathbb{R}$ satisfies $\partial f(x_0) = \{\nabla f(x_0)\}$ for every $x_0 \in U$ i.e., there is a unique subgradient (the gradient!) at each point (see Exercise 9.6).

For convex functions, which are not necessarily differentiable, there may be several subgradients at a point. This is similar to the fact there there can be several supporting hyperplanes at a boundary point of a convex subset. The mother of all examples is $f(x) = |x|$. Here

$$\partial f(0) = [-1, 1]$$

whereas $\partial f(-1) = \{-1\}$ and $\partial f(1) = \{1\}$. In general we have the following result.

Theorem 9.3. *Let* $f : U \to \mathbb{R}$ *be a function, where* U *is an open convex subset of* $\mathbb{R}^n$. *Then* f *is convex if and only if* $\partial f(x) \neq \emptyset$ *for every* $x \in U$.

Proof.   Suppose that $\partial f(x) \neq \emptyset$ for every $x \in U$. Then the proof of the first part of Theorem 9.1 carries over verbatim showing that f is a convex function. Suppose on the other hand that f is a convex function and let $x_0 \in U$. We wish to prove that $\partial f(x_0) \neq \emptyset$. The epigraph

$$\mathrm{epi}(f) = \{(x, y) \mid x \in U, y \in \mathbb{R}, f(x) \leq y\}$$

is a convex subset of $\mathbb{R}^{n+1}$ by Lemma 7.2. The point $w = (x_0, f(x_0)) \in \mathrm{epi}(f)$ has to belong to the boundary $\partial \mathrm{epi}(f)$, since $(x_0, f(x_0) - \epsilon) \notin \mathrm{epi}(f)$ for every $\epsilon > 0$. Therefore there exists a supporting hyperplane H for $\mathrm{epi}(f)$ at w by Theorem 6.11. Suppose H is given as the set of points $(x, y) \in \mathbb{R}^{n+1}$ with

$$(\alpha_1, \alpha_2)^t(x, y) = \beta,$$

for $\alpha_1, x \in \mathbb{R}^n$ and $\alpha_2, y, \beta \in \mathbb{R}$. Since $(x_0, f(x_0)) \in H$,

$$\beta = (\alpha_1, \alpha_2)^t(x_0, f(x_0)).$$

Also, as $(\alpha_1, \alpha_2)^t(x, y) \geq \beta$ for every $(x, y) \in \mathrm{epi}(f)$ we have

$$\alpha_1^t(x - x_0) + \alpha_2(y - f(x_0)) \geq 0 \tag{9.3}$$

for every $x \in U$ and $y \geq f(x)$. Inserting (x_0, y) with $y > f(x_0)$ in (9.3), it follows that $\alpha_2 \geq 0$. If $\alpha_2 > 0$ we may divide (9.3) by α_2 to get

$$\alpha_1^t / \alpha_2(x - x_0) + y - f(x_0) \geq 0. \tag{9.4}$$

Putting $y = f(x)$ in (9.4), we see that $\xi = -\alpha_1/\alpha_2 \in \partial f(x_0)$ is a subgradient at x_0.

    We go on to prove $\alpha_2 > 0$. Since U is open and $x_0 \in U$ we may find $\epsilon > 0$ such that

$$|x - x_0| < \epsilon \Rightarrow x \in U. \tag{9.5}$$

If $\alpha_1 \neq 0$ then $\alpha_1^t(x - x_0)$ can always be made negative for suitable choice of x in (9.5). In this case we must have $\alpha_2 > 0$ for (9.3) to hold. If $\alpha_1 = 0$ we must have $\alpha_2 > 0$, since a supporting hyperplane cannot have both $\alpha_1 = 0$ and $\alpha_2 = 0$. Therefore $\alpha_2 > 0$ and the proof is complete. $\square$

9.2   Convexity and the Hessian

Let $U \subseteq \mathbb{R}^n$ be an open subset and $f : U \to \mathbb{R}$ a differentiable function with continuous second order partial derivatives (see Theorem 8.5). We wish to analyze how f behaves around a given $x_0 \in U$. A natural way of doing this is studying the one variable function

$$f_d(t) = f(x_0 + td)$$

for some direction vector $d \in \mathbb{R}^n$. Since U is an open subset, it follows that $x_0 + td \in U$ for $t \in (-\delta, \delta)$ where δ is positive and small. Notice that x_0 is a local minimum (or maximum) for f if and only if 0 is a local minimum (or maximum) for f_d for every $d \in \mathbb{R}^n$. A measure of growth of f at x_0 in the direction of d is $f_d'(0)$. The chain rule shows that

$$f_d'(0) = \nabla f(x_0)d.$$

If x_0 is a critical point for f, then $f_d'(0) = 0$ for every $d \in \mathbb{R}^n$. We can use Theorem 7.18 to investigate if x_0 is a local minimum by computing $f_d''(0)$. The chain rule shows

$$f_d'(t) = \nabla f(x_0 + td)d. \tag{9.6}$$

In Exercise 9.4 you are asked to dig deeper into the chain rule and show that

$$f_d''(t) = (f_d')'(t) = d^t \nabla^2 f(x_0 + td)d,$$

where

$$\nabla^2 f(x) := \begin{pmatrix} \dfrac{\partial^2 f}{\partial x_1 \partial x_1}(x) & \cdots & \dfrac{\partial^2 f}{\partial x_1 \partial x_n}(x) \\ \vdots & \ddots & \vdots \\ \dfrac{\partial^2 f}{\partial x_n \partial x_1}(x) & \cdots & \dfrac{\partial^2 f}{\partial x_n \partial x_n}(x) \end{pmatrix}.$$

The matrix $\nabla^2 f(x)$ is called *the Hessian matrix*[1] or simply the *Hessian* of f at x. Using Theorem 7.18 in telling whether a critical point is a local minimum or maximum, we therefore focus on

$$f_d''(0) = d^t \nabla^2 f(x_0)d, \tag{9.7}$$

for $d \in \mathbb{R}^n$. Notice that (9.7) is gotten as $v^t Av$, where A is a quadratic matrix and $v \in \mathbb{R}^n$ is a vector. Recall that a matrix A is called *symmetric* if $A = A^t$. Under our assumptions, the Hessian of f at a point is a symmetric matrix by Theorem 8.5. Now the following definition is very natural.

---

[1] Otto Hesse (1811–1874). German mathematician.

Definition 9.4. A real symmetric matrix A is called *positive definite* if

$$v^t A v > 0$$

for every $v \in \mathbb{R}^n \setminus \{0\}$ and *positive semidefinite*

$$v^t A v \geq 0$$

for every $v \in \mathbb{R}^n$.

Before reading on, you should experiment a bit with Definition 9.4 if you have only seen it briefly in the context of a standard calculus or linear algebra course. Give at least examples (only using Definition 9.4) of symmetric 2×2 matrices being

(1) not positive semidefinite
(2) non-zero positive semidefinite but not positive definite
(3) positive definite but not a diagonal matrix.

Here is the multivariate generalization of Corollary 7.20.

Theorem 9.5. *Let $f : U \to \mathbb{R}$ be a differentiable function with continuous second order partial derivatives, where $U \subseteq \mathbb{R}^n$ is a convex open subset. Then f is convex if and only if the Hessian $\nabla^2 f(x)$ is positive semidefinite for every $x \in U$. If $\nabla^2 f(x)$ is positive definite for every $x \in U$, then f is strictly convex.*

Proof.   We have done all the work for a convenient reduction to the one variable case. Suppose that f is convex. Then the same reasoning as in the proof of Theorem 9.1 shows that

$$g(t) = f(x + tv)$$

is a convex function for every $x \in U$ and every $v \in \mathbb{R}^n$ from an open interval $(-\delta, \delta)$ to $\mathbb{R}$ for suitable $\delta > 0$. Therefore $g''(0) = v^t \nabla^2 f(x) v \geq 0$ by Theorem 7.19. This proves that the matrix $\nabla^2 f(x)$ is positive semidefinite for every $x \in U$. Suppose on the other hand that $\nabla^2 f(x)$ is positive semidefinite for every $x \in U$. Then Theorem 7.19 shows that $g(t) = f(x + t(y - x))$ is a convex function from $(-\delta, 1 + \delta)$ to $\mathbb{R}$ for $\delta > 0$ small and $x, y \in U$, since

$$g''(\alpha) = (y - x)^t \nabla^2 f(x + \alpha(y - x))(y - x) \geq 0$$

for $0 \leq \alpha \leq 1$. Therefore f is a convex function, since

$$f((1-t)x + ty) = g((1-t) \cdot 0 + t \cdot 1)$$
$$\leq (1-t)g(0) + tg(1) = (1-t)f(x) + tf(y).$$

The same argument (using the last part of Theorem 7.19 on strict convexity), shows that g is strictly convex if $\nabla^2 f(x)$ is positive definite. It follows that f is strictly convex if $\nabla^2 f(x)$ is positive definite for every $x \in U$. $\square$

We move on to study positive definite and positive semidefinite matrices in some more detail.

9.3 Positive definite and positive semidefinite matrices

Lemma 9.6. *A diagonal matrix*

$$D = \begin{pmatrix} \lambda_1 & 0 & \cdots & 0 \\ 0 & \lambda_2 & \cdots & 0 \\ \vdots & \vdots & \ddots & \vdots \\ 0 & 0 & \cdots & \lambda_n \end{pmatrix}$$

is positive definite if and only if $\lambda_1, \ldots, \lambda_n > 0$ and positive semidefinite if and only if $\lambda_1, \ldots, \lambda_n \geq 0$. If B is an invertible matrix and A is a symmetric matrix, then A is positive definite (semidefinite) if and only if $B^t A B$ is positive definite (semidefinite).

Proof. The rules of matrix multiplication show that

$$v^t D v = \lambda_1 x_1^2 + \cdots + \lambda_n x_n^2,$$

where $v = (x_1, \ldots, x_n)^t \in \mathbb{R}^n$. This shows that D is positive definite if and only if $\lambda_1, \ldots, \lambda_n > 0$ and similarly positive semidefinite if and only if $\lambda_1, \ldots, \lambda_n \geq 0$ using Definition 9.4.

Recall that $(AB)^t = B^t A^t$ for matrices, where the multiplication makes sense. If B is invertible, then $v \neq 0$ if and only if $Bv \neq 0$ and since

$$v^t(B^t A B)v = (Bv)^t A(Bv)$$

it follows that $B^t A B$ is positive definite (semidefinite) if A is positive definite (semidefinite). Since $(B^t)^{-1} = (B^{-1})^t$ we have

$$v^t A v = (B^{-1}v)^t B^t A B(B^{-1}v)$$

and it follows that A is positive definite (semidefinite) if $B^t A B$ is positive definite (semidefinite). $\qquad\square$

Lemma 9.6 is at the heart of a very useful method for deciding if a given symmetric matrix A is positive semidefinite. The key point is that we can always find an invertible matrix B, such that $B^t A B$ is diagonal. Consider as a small example, the symmetric 2×2 matrix

$$A = \begin{pmatrix} 1 & 2 \\ 2 & 1 \end{pmatrix}.$$

Multiplying A from the right by

$$B = \begin{pmatrix} 1 & -2 \\ 0 & 1 \end{pmatrix}$$

corresponds to the elementary column operation adding -2 times the first column to the second column. Similarly multiplying B^t from the left corresponds to the elementary row operation adding -2 times the first row to the second row. All in all, we get

$$B^t A B = \begin{pmatrix} 1 & 0 \\ 0 & -3 \end{pmatrix}$$

showing by Lemma 9.6 that A is not positive semidefinite. The next example shows how Lemma 9.6 applies to general symmetric 2×2 matrices.

Example 9.7. Let

$$A = \begin{pmatrix} a & c \\ c & b \end{pmatrix}$$

be a symmetric 2×2 matrix. If A is positive definite, we must have $a > 0$, since

$$(1, 0) A \begin{pmatrix} 1 \\ 0 \end{pmatrix} = a.$$

In fact A is positive definite if and only if $a > 0$ and $ab - c^2 > 0$. This follows from Lemma 9.6 by defining

$$B = \begin{pmatrix} 1 & -\frac{c}{a} \\ 0 & 1 \end{pmatrix}$$

and observing that

$$B^t A B = \begin{pmatrix} a & 0 \\ 0 & b - \frac{c^2}{a} \end{pmatrix}.$$

The following example shows how our efforts in this chapter apply in deducing the celebrated least squares method of Gauss in the plane i.e., given a data set $(x_1, y_1), \ldots, (x_n, y_n) \in \mathbb{R}^2$ we wish to compute the best approximating line $y = ax + \beta$.

Example 9.8. Consider the three points $(1, 2)$, $(2, 1)$ and $(4, 3)$ in $\mathbb{R}^2$. What line $y = ax + b$ fits these points best? If best fit is interpreted as minimizing the maximal vertical distance (the maximal error)

$$\max\{|2 - a - b|, |1 - 2a - b|, |3 - 4a - b|\}$$

to the points, then the problem can be solved using linear programming as explained in Example 5.16.

The famous least squares method of Gauss computes the best fit interpreted as minimizing the sum of squared distances to the points i.e., a and b are chosen to minimize

$$(2 - a - b)^2 + (1 - 2a - b)^2 + (3 - 4a - b)^2.$$

We will study the least squares method from the point of view of convex functions.

Consider finitely many points $(x_1, y_1), \ldots, (x_n, y_n)$ in $\mathbb{R}^2$ and the function

$$f(a, b) = \sum_{j=1}^{n} (y_j - ax_j - b)^2.$$

Then $f : \mathbb{R}^2 \to \mathbb{R}$ is a non-negative differentiable function. We will compute the Hessian of f. Notice first that

$$\frac{\partial f}{\partial a} = -2 \sum_{j=1}^{n} x_j(y_j - ax_j - b)$$

$$\frac{\partial f}{\partial b} = -2 \sum_{j=1}^{n} y_j - ax_j - b \tag{9.8}$$

and furthermore

$$\frac{\partial^2 f}{\partial a^2} = 2 \sum_{j=1}^{n} x_j^2, \quad \frac{\partial^2 f}{\partial a \partial b} = 2 \sum_{j=1}^{n} x_j \quad \text{and} \quad \frac{\partial^2 f}{\partial b^2} = 2n.$$

Therefore

$$\nabla^2 f(a, b) = \begin{pmatrix} 2 \sum_{j=1}^{n} x_j^2 & 2 \sum_{j=1}^{n} x_j \\ 2 \sum_{j=1}^{n} x_j & 2n \end{pmatrix}$$

and the function f is strictly convex by Theorem 9.5 and Example 9.7 if

$$4n \sum_{j=1}^{n} x_j^2 > 4 \Big(\sum_{j=1}^{n} x_j \Big)^2.$$

This inequality can be verified by the *Cauchy-Schwarz inequality* (see Lemma A.1) for the vectors $u = (1, 1, \ldots, 1)$ and x in $\mathbb{R}^n$ i.e.,

$$|u|^2 |x|^2 \geq (u^t x)^2$$

with equality if and only if $u = \lambda x$ for some $\lambda \in \mathbb{R}$. So if $x_i \neq x_j$ for some $i \neq j$, then f is strictly convex. In this case we investigate the function for a (unique) global minimum using Theorem 9.1. We get by $\nabla f(a, b) = 0$ in (9.8) the following system of linear equations in a and b

$$\Big(\sum_{j=1}^{n} x_j^2 \Big) a + \Big(\sum_{j=1}^{n} x_j \Big) b = \sum_{j=1}^{n} x_j y_j$$

$$\Big(\sum_{j=1}^{n} x_j \Big) a + nb = \sum_{j=1}^{n} y_j$$

giving

$$a = \frac{nC_1 - C_2 A_2}{nA_1 - A_2^2} \quad \text{and} \quad b = \frac{A_1 C_2 - A_2 C_1}{nA_1 - A_2^2}$$

with

$$A_1 = \sum_{j=1}^{n} x_j^2, \quad A_2 = \sum_{j=1}^{n} x_j, \quad C_1 = \sum_{j=1}^{n} x_j y_j \quad \text{and} \quad C_2 = \sum_{j=1}^{n} y_j$$

by *Cramer's formula*. In our example the best least squares approximating line turns out to be $y = \frac{3}{7}x + 1$. In case you wonder, the best approximating line when the maximal error must be small, is $y = \frac{1}{3}x + 1$. This line has maximal error $\frac{2}{3}$, whereas the least squares line has maximal error $\frac{6}{7}$.

9.4   Principal minors and definite matrices

Definition 9.9.   Let A be an $n \times n$ matrix and $I, J \subseteq [n]$ with $[n] = \{1, \ldots, n\}$. By $A[I, J]$ we denote the *submatrix* with rows from I and columns from J. If $|I| = |J|$ we call $\det A[I, J]$ the *minor* given by I and J. If $I = J$, $A[I, J]$ is called a *principal submatrix*. The principal submatrices given by $I = [1], [2], \ldots, [n]$ are called the *leading principal submatrices*. A *(leading) principal minor* is defined as the determinant of a (leading) principal submatrix.

Example 9.10. Suppose that

$$A = \begin{pmatrix} 1 & 5 & 2 & 1 \\ 5 & 3 & 0 & 2 \\ 2 & 0 & 5 & 3 \\ 1 & 2 & 3 & 4 \end{pmatrix},$$

$I = [2] = \{1, 2\}$ and $J = \{1, 3, 4\}$. Then

$$A[J, J] = \begin{pmatrix} 1 & 2 & 1 \\ 2 & 5 & 3 \\ 1 & 3 & 4 \end{pmatrix}, \quad A[I, J] = \begin{pmatrix} 1 & 2 & 1 \\ 5 & 0 & 2 \end{pmatrix}, \quad A[I, I] = \begin{pmatrix} 1 & 5 \\ 5 & 3 \end{pmatrix}.$$

Here $A[J, J]$ is a principal submatrix and $A[I, I]$ a leading principal submatrix.

For an $n \times n$ matrix A, we will use the notation $A[r]$ for the r-th leading principal minor $A[[r], [r]]$, where $1 \leq r \leq n$. Here is the generalization of Example 9.7 to symmetric $n \times n$ matrices.

Theorem 9.11. *A symmetric $n \times n$ matrix $A = (a_{ij})_{1 \leq i, j \leq n}$ is positive definite if and only if its leading principal minors are > 0.*

Proof. This proof is really a souped up version of Example 9.7. Make sure you understand that example before reading this proof. We will prove the statement by induction on n, the case $n = 1$ reading $a_{11}x^2 > 0$ for every $x \neq 0$ if and only if $a_{11} > 0$. In the general case it is no restriction (why?) to assume that $a_{11} > 0$. With this assumption we let

$$B = \begin{pmatrix} 1 & -\frac{a_{12}}{a_{11}} & \cdots & -\frac{a_{1n}}{a_{11}} \\ 0 & 1 & \cdots & 0 \\ \vdots & \vdots & \ddots & \vdots \\ 0 & 0 & \cdots & 1 \end{pmatrix}.$$

As in Example 9.7 we get

$$B^t A B = \begin{pmatrix} a_{11} & 0 & \cdots & 0 \\ 0 & c_{11} & \cdots & c_{1,n-1} \\ \vdots & \vdots & \ddots & \vdots \\ 0 & c_{n-1,1} & \cdots & c_{n-1,n-1} \end{pmatrix}, \tag{9.9}$$

where $C = (c_{ij})$, $1 \leq i, j \leq n - 1$ is a symmetric $(n - 1) \times (n - 1)$ matrix. Therefore A is positive definite if and only if C is positive definite (Exercise 9.21). However, for the principal $r \times r$ minors we have

$$\det A[r] = \det(B^t AB)[r] = a_{11} \det C[r - 1] \qquad (9.10)$$

for $2 \leq r \leq n$, since the determinant is invariant under elementary row- and column operations. If $\det A[r] > 0$ for every $r = 2, \ldots, n$, then $\det C[k] > 0$ for every $k = 1, \ldots, n - 1$ by (9.10) and it follows by induction that C is positive definite. Therefore A is a positive definite. On the other hand, if A is positive definite, then C is positive definite and the induction assumption gives that $\det C[k] > 0$ for every $k = 1, \ldots, n - 1$ and (9.10) shows that $\det A[r] > 0$ for every $r = 1, \ldots, n$. □

9.5 The positive semidefinite cone

The set $\mathcal{S}$ of symmetric $n \times n$ matrices may be viewed as the vector space $\mathbb{R}^{n(n+1)/2}$. The subset of positive semidefinite matrices is a closed convex cone in $\mathcal{S}$. These facts are a consequence of Definition 9.4 (see Exercise 9.25 and also Exercise 9.27 for a description of the extreme rays in $\mathcal{S}$). We wish to give a criterion for a symmetric matrix to be positive semidefinite analogously to Theorem 9.11. Intuitively we need to describe the closure of the set of positive definite matrices in $\mathcal{S}$.

Example 9.12. A symmetric 2×2 matrix

$$A = \begin{pmatrix} a & c \\ c & b \end{pmatrix}$$

is positive definite if and only if $a > 0$ and $ab - c^2 > 0$. It is tempting to guess that A is positive semidefinite if and only if $a \geq 0$ and $ab - c^2 \geq 0$. The matrix

$$A = \begin{pmatrix} 0 & 0 \\ 0 & -1 \end{pmatrix}$$

shows that this is wrong. In mathematics, the closure of a subset is a rather subtle operation. In our setting, more inequalities are needed to capture positive semidefiniteness.

Our starting point is the following.

Lemma 9.13. *Let $A \in \mathcal{S}$. Then A is positive semidefinite if and only if*

$$A + \epsilon I$$

is positive definite for every $\epsilon > 0$, where $I \in \mathcal{S}$ is the identity matrix.

Proof. This follows from Definition 9.4, since for every $x \in \mathbb{R}^n \setminus \{0\}$ we have

$$x^t A x \geq 0 \iff x^t(A + \epsilon I)x = x^t A x + \epsilon|x|^2 > 0$$

for every $\epsilon > 0$. □

The above lemma naturally leads us to study the function f defined below.

Proposition 9.14. *Let A be an $n \times n$ matrix and let $f : \mathbb{R} \to \mathbb{R}$ be given by*

$$f(t) = \det(A + tI_n),$$

where I_n is the identity $n \times n$ matrix. Then f is differentiable and

$$f'(t) = \det(A^{(1)} + tI_{n-1}) + \cdots + \det(A^{(n)} + tI_{n-1}),$$

where $A^{(i)}$ is the $(n-1) \times (n-1)$ matrix with the i-th row and i-th column removed i.e., the principal minor $A[I, I]$, where $I = \{1, \ldots, n\} \setminus \{i\}$.

Proof. Let M_n denote the set of $n \times n$ matrices identified with $\mathbb{R}^{n^2}$. The determinant is a differentiable function $\det : M_n \to \mathbb{R}$. For $X = (x_{ij}) \in M_n$, expansion of the determinant from the i-th row gives

$$\det X = x_{i1}D_{i1} + \cdots + x_{ij}D_{ij} + \cdots + x_{in}D_{in},$$

where D_{rs} is $(-1)^{r+s}$ times the determinant of X with the r-th row and s-th column removed. This fact from linear algebra shows that the gradient of the determinant is

$$\nabla \det(X) = \left(\frac{\partial \det}{\partial x_{ij}}\right)_{1 \leq i,j \leq n} = (D_{ij})_{1 \leq i,j \leq n}.$$

The function $g : \mathbb{R} \to M_n$ given by $g(t) = A + tI$ is differentiable with $g'(t) = I_n$. Since $f = \det \circ g$, we get by the chain rule that f is differentiable with

$$\begin{aligned} f'(t) &= \nabla \det(g(t))g'(t) = \nabla \det(A + tI_n)I_n \\ &= \det(A^{(1)} + tI_{n-1}) + \cdots + \det(A^{(n)} + tI_{n-1}), \end{aligned}$$

where the product $\nabla \det(A + tI_n)I_n$ is interpreted as the dot product on each of the entries in M_n. □

Usually the following result is proved using the spectral theorem (see §9.7). I owe the following clever proof to Tage Bai Andersen.

Theorem 9.15. *A symmetric $n \times n$ matrix $A = (a_{ij})_{1 \leq i,j \leq n}$ is positive semidefinite if and only if its $2^n - 1$ principal minors are non-negative.*

Proof.   Let $e_1, \ldots, e_n$ denote the canonical basis vectors of $\mathbb{R}^n$. Suppose A is positive semidefinite and let $J = \{i_1, \ldots, i_r\} \subseteq [n]$. Then

$$v^t A v = u^t A[J, J] u$$

for $v \in \mathbb{R}^J$ and $u = (x_{i_1}, \ldots, x_{i_r})^t \in \mathbb{R}^r$, where $v = x_{i_1} e_{i_1} + \cdots + x_{i_r} e_{i_r}$. Therefore the symmetric $r \times r$ matrix $A[J, J]$ is positive semidefinite and $A[J, J] + \epsilon I$ is positive definite for every $\epsilon > 0$ by Lemma 9.13. By Theorem 9.11,

$$\det(A[J, J] + \epsilon I) > 0$$

for every $\epsilon > 0$. This implies $\det(A[J, J]) \geq 0$, since $f(t) = \det(A[J, J] + tI)$ is a continuous function by Proposition 9.14.

On the other hand suppose that $\det(A[J, J]) \geq 0$ for every subset $J \subseteq [n]$. We will prove that A is positive semidefinite by induction on n. For $n = 1$ it is true. If A is an $n \times n$ matrix for $n > 1$, it suffices by Lemma 9.13 to prove that $A + \epsilon I$ is positive definite for every $\epsilon > 0$. Using Theorem 9.11, this can be done by showing that the leading principal minors of $A + \epsilon I$ are positive. For this, let

$$f_m(t) = \det(A[m] + tI_m),$$

where $m = 1, \ldots, n$. By Proposition 9.14 and induction it follows that $f_m'(t) > 0$ for $t > 0$. Therefore $f_m(t) > 0$ for $t > 0$, since $f_m(0) \geq 0$. This shows that the m-th leading principal minor of $A + \epsilon I_n$ is positive for $\epsilon > 0$ finishing the proof.   □

Example 9.16. Here are a few examples of Theorem 9.15 in action. The 2×2 symmetric matrix

$$\begin{pmatrix} a & c \\ c & b \end{pmatrix}$$

is positive semidefinite if and only if its principal minors

$$a \geq 0 \qquad \{1\}$$
$$b \geq 0 \qquad \{2\}$$
$$ab - c^2 \geq 0 \qquad \{1, 2\}$$

are non-negative. Comparing this to Example 9.12 we see that $b \geq 0$ was the missing inequality. The 3×3 symmetric matrix

$$
\begin{pmatrix}
a & d & e \\
d & b & f \\
e & f & c
\end{pmatrix}
$$

is positive semidefinite if and only if its principal minors

$$
\begin{aligned}
a &\geq 0 & \{1\} \\
b &\geq 0 & \{2\} \\
c &\geq 0 & \{3\} \\
ab - d^2 &\geq 0 & \{1,2\} \\
ac - e^2 &\geq 0 & \{1,3\} \\
bc - f^2 &\geq 0 & \{2,3\} \\
abc + 2dfe - af^2 - be^2 - cd^2 &\geq 0 & \{1,2,3\}
\end{aligned}
$$

are non-negative.

9.6 Reduction of symmetric matrices

For a 17×17 symmetric real matrix A, you need to verify $2^{17} - 1 = 131\,071$ inequalities in general using Theorem 9.15 to show that A is positive semidefinite. There is much faster way of deciding this using Lemma 9.6. The key is the reduction of a symmetric matrix to a diagonal matrix.

Theorem 9.17. *Let A be a real symmetric $n \times n$ matrix. Then there exists an invertible matrix B, such that $B^t A B$ is a diagonal matrix.*

Proof. Suppose that $A = (a_{ij})$. If A has a non-zero entry in the upper left hand corner i.e., $a_{11} \neq 0$, then

$$
B_1^t A B_1 =
\begin{pmatrix}
a_{11} & 0 & \cdots & 0 \\
0 & c_{11} & \cdots & c_{1,n-1} \\
\vdots & \vdots & \ddots & \vdots \\
0 & c_{n-1,1} & \cdots & c_{n-1,n-1},
\end{pmatrix}
$$

where $C = (c_{ij})$ is a real symmetric matrix and B_1 is the invertible $n \times n$

matrix

$$\begin{pmatrix} 1 & -\dfrac{a_{12}}{a_{11}} & \cdots & -\dfrac{a_{1n}}{a_{11}} \\ 0 & 1 & \cdots & 0 \\ \vdots & \vdots & \ddots & \vdots \\ 0 & 0 & \cdots & 1 \end{pmatrix}.$$

By induction on n we may find an invertible matrix $(n-1) \times (n-1)$ matrix B_2 such that

$$B_2^t C B_2 = \begin{pmatrix} a_1 & 0 & \cdots & 0 \\ 0 & a_2 & \cdots & 0 \\ \vdots & \vdots & \ddots & \vdots \\ 0 & 0 & \cdots & a_{n-1} \end{pmatrix}.$$

Putting

$$B = B_1 \begin{pmatrix} 1 & 0 \\ 0 & B_2 \end{pmatrix}$$

it follows that

$$B^t A B = \begin{pmatrix} a_{11} & 0 & \cdots & 0 \\ 0 & a_1 & \cdots & 0 \\ \vdots & \vdots & \ddots & \vdots \\ 0 & 0 & \cdots & a_{n-1} \end{pmatrix}.$$

We now treat the case of a zero entry in the upper left hand corner i.e., $a_{11} = 0$. Suppose first that $a_{jj} \neq 0$ for some $j > 1$. Let P denote the identity matrix with the first and j-th rows interchanged. The operation $A \mapsto AP$ amounts to interchanging the first and j-th columns in A. Similarly $A \mapsto P^t A$ is interchanging that first and j-th rows in A. The matrix P is invertible and $P^t A P$ is a symmetric matrix with $(P^t A P)_{11} = a_{jj} \neq 0$ and we have reduced to the case of a non-zero entry in the upper left hand corner.

If $a_{ii} = 0$ for every $i = 1, \ldots, n$ we may assume that $a_{1j} \neq 0$ for some $j > 1$. Let B denote the identity matrix where the entry in the first column and j-th row is 1. The operation $A \mapsto AB$ amounts to adding the j-th column to the first column in A. Similarly $A \mapsto B^t A$ is adding the j-th row to the first row in A. All in all we get $(B^t A B)_{11} = 2a_{1j} \neq 0$, where we have used that $a_{ii} = 0$ for $i = 1, \ldots, n$. Again we have reduced to the case of a non-zero entry in the upper left hand corner. $\square$

The three reduction steps in the proof of Theorem 9.17 are illustrated using actual numbers in the Examples 9.19, 9.20 and 9.21. Let us first record the following corollary, which in a sense says that positive semidefinite matrices are the ones that admit a square root.

Corollary 9.18. *Let A be a real symmetric $n \times n$ matrix. Then there exists a diagonal matrix D and an invertible matrix C, such that*

$$A = C^t DC. \tag{9.11}$$

A is positive semidefinite if and only if there exists a matrix S such that $A = S^t S$. Every positive semidefinite matrix is a sum of at most n positive semidefinite rank one matrices.

Proof. If A is symmetric, Theorem 9.17 gives the existence of an invertible matrix B such that $B^t AB$ is a diagonal matrix D. Putting $C = B^{-1}$ shows $A = C^t DC$.

If A is positive semidefinite, then the entries in D are non-negative. Let $\sqrt{D}$ denote the matrix, where the entries are the positive square roots of the entries in D. Then $D = \sqrt{D}^t \sqrt{D}$ and $A = C^t DC = (\sqrt{D}C)^t(\sqrt{D}C)$. It is left as an exercise to show that any matrix of the form $S^t S$, where S is an $m \times n$ matrix is positive semidefinite (Exercise 9.19). The last result follows from the decomposition $A = S^t S$ by the identity

$$S^t S = v_1 v_1^t + \cdots + v_n v_n^t,$$

where $v_1, \ldots, v_n$ are the row vectors of S. It is not too hard to verify that the matrix uu^t is a symmetric matrix of rank one, where u is any non-zero vector (Exercise 9.19). $\qquad\square$

Example 9.19. Consider the 3×3 real symmetric matrix.

$$A = (a_{ij}) = \begin{pmatrix} 1 & 5 & 2 \\ 5 & 3 & 0 \\ 2 & 0 & 5 \end{pmatrix}.$$

Here $a_{11} = 1 \neq 0$. Therefore the fundamental step in the proof of Theorem 9.17 applies and

$$\begin{pmatrix} 1 & 0 & 0 \\ -5 & 1 & 0 \\ -2 & 0 & 1 \end{pmatrix} A \begin{pmatrix} 1 & -5 & -2 \\ 0 & 1 & 0 \\ 0 & 0 & 1 \end{pmatrix} = \begin{pmatrix} 1 & 0 & 0 \\ 0 & -22 & -10 \\ 0 & -10 & 1 \end{pmatrix}$$

and again

$$\begin{pmatrix} 1 & 0 & 0 \\ 0 & 1 & 0 \\ 0 & -\frac{5}{11} & 1 \end{pmatrix} \begin{pmatrix} 1 & 0 & 0 \\ 0 & -22 & -10 \\ 0 & -10 & 1 \end{pmatrix} \begin{pmatrix} 1 & 0 & 0 \\ 0 & 1 & -\frac{5}{11} \\ 0 & 0 & 1 \end{pmatrix} = \begin{pmatrix} 1 & 0 & 0 \\ 0 & -22 & 0 \\ 0 & 0 & \frac{61}{11} \end{pmatrix}.$$

Summing up we get

$$B = \begin{pmatrix} 1 & -5 & -2 \\ 0 & 1 & 0 \\ 0 & 0 & 1 \end{pmatrix} \begin{pmatrix} 1 & 0 & 0 \\ 0 & 1 & -\frac{5}{11} \\ 0 & 0 & 1 \end{pmatrix} = \begin{pmatrix} 1 & -5 & \frac{3}{11} \\ 0 & 1 & -\frac{5}{11} \\ 0 & 0 & 1 \end{pmatrix}.$$

You are invited to check that

$$B^t A B = \begin{pmatrix} 1 & 0 & 0 \\ 0 & -22 & 0 \\ 0 & 0 & \frac{61}{11} \end{pmatrix}.$$

Example 9.20. Let

$$A = \begin{pmatrix} 0 & 0 & 1 & 1 \\ 0 & 0 & 2 & 3 \\ 1 & 2 & 1 & 4 \\ 1 & 3 & 4 & 0 \end{pmatrix}.$$

Here $a_{11} = a_{22} = 0$, but the diagonal element $a_{33} \neq 0$. So we are in the second step of the proof of Theorem 9.17. Using the matrix

$$P = \begin{pmatrix} 0 & 0 & 1 & 0 \\ 0 & 1 & 0 & 0 \\ 1 & 0 & 0 & 0 \\ 0 & 0 & 0 & 1 \end{pmatrix}$$

we get

$$P^t A P = \begin{pmatrix} 1 & 2 & 1 & 4 \\ 2 & 0 & 0 & 3 \\ 1 & 0 & 0 & 1 \\ 4 & 3 & 1 & 0 \end{pmatrix}.$$

As argued in the proof, this corresponds to interchanging the first and third columns and then interchanging the first and third rows. In total you move the non-zero a_{33} to the upper left corner in the matrix.

Example 9.21. Consider the symmetric matrix

$$A = \begin{pmatrix} 0 & 1 & 1 & 1 \\ 1 & 0 & 1 & 1 \\ 1 & 1 & 0 & 1 \\ 1 & 1 & 1 & 0 \end{pmatrix}.$$

We have zero entries in the diagonal. As in the third step in the proof of Theorem 9.17 we must find an invertible matrix B_1, such that the upper left corner in $B_1^t A B_1$ is non-zero. In the proof it is used that every diagonal element is zero: if we locate a non-zero element in the j-th column in the first row, we can add the j-th column to the first column and then the j-th row to the first row obtaining a non-zero element in the upper left corner. For A above we choose $j = 2$ and the matrix B_1 becomes

$$B_1 = \begin{pmatrix} 1 & 0 & 0 & 0 \\ 1 & 1 & 0 & 0 \\ 0 & 0 & 1 & 0 \\ 0 & 0 & 0 & 1 \end{pmatrix}$$

so that

$$B_1^t A B_1 = \begin{pmatrix} 2 & 1 & 2 & 2 \\ 1 & 0 & 1 & 1 \\ 2 & 1 & 0 & 1 \\ 2 & 1 & 1 & 0 \end{pmatrix}.$$

9.7 The spectral theorem

Recall that a number $\lambda \in \mathbb{R}$ is called an eigenvalue for a real $n \times n$ matrix A if there exists a non-zero vector $v \in \mathbb{R}^n$ with $Av = \lambda v$. Such a vector is called an eigenvector for λ and it satisfies $(A - \lambda I)v = 0$, where I is the $n \times n$ identity matrix.

If $\mathbb{R}^n$ has a basis of eigenvectors $v_1, \ldots, v_n$ with eigenvalues $\lambda_1, \ldots, \lambda_n$ for A, then

$$T^{-1}AT = \begin{pmatrix} \lambda_1 & 0 & \cdots & 0 \\ 0 & \lambda_2 & \cdots & 0 \\ \vdots & \vdots & \ddots & \vdots \\ 0 & 0 & \cdots & \lambda_n \end{pmatrix}, \tag{9.12}$$

where T is the (invertible) matrix with columns $v_1, \ldots, v_n$ i.e., $Te_i = v_i$ with $e_1, \ldots, e_n$ the standard basis of $\mathbb{R}^n$ (and therefore $T^{-1}v_i = e_i$).

Example 9.22. The matrix

$$A = \begin{pmatrix} 1 & 1 \\ 0 & 1 \end{pmatrix}$$

is an example of a matrix not having a basis of eigenvectors. The only possible eigenvalue for A is $\lambda = 1$, since $A - \lambda I$ is invertible for $\lambda \neq 1$. For $\lambda = 1$ we have

$$\left(\begin{pmatrix} 1 & 1 \\ 0 & 1 \end{pmatrix} - \begin{pmatrix} 1 & 0 \\ 0 & 1 \end{pmatrix} \right) v = \begin{pmatrix} 0 & 1 \\ 0 & 0 \end{pmatrix} v = 0.$$

This is satisfied only for vectors of the form $v = ce_2$, where $c \in \mathbb{R}^n$. No such two vectors can be a basis of $\mathbb{R}^2$.

The identity (9.12) is similar to Theorem 9.17 if T satisfies $T^{-1} = T^t$. This (surprisingly) happens for symmetric matrices and it is called the spectral theorem.

Theorem 9.23. *Let A be a real symmetric matrix. Then there exists an invertible matrix T with $T^{-1} = T^t$, such that $T^t A T$ is a diagonal matrix (having the eigenvalues of A as diagonal entries).*

Not every real matrix has (real) eigenvalues. Consider as a simple example that 2×2 matrix

$$A = \begin{pmatrix} 1 & 1 \\ -1 & 1 \end{pmatrix}.$$

If $Av = \lambda v$ for some non-zero v, then you can check that $\lambda^2 - 2\lambda + 2 = (\lambda - 1)^2 + 1 = 0$. This is impossible if $\lambda \in \mathbb{R}$. If A is a symmetric matrix, a small miracle happens.

Theorem 9.24. *Let A be a real symmetric matrix. Then A has a real eigenvalue.*

Proof.   This proof goes back to 1826 and is due to Cauchy (Figure 9.1). Suppose that A is a real symmetric $n \times n$ matrix and consider the optimization problem

$$M = \max\{x^t A x \mid x \in S^n\},$$

where $S^n = \{x \in \mathbb{R}^n \mid x^t x = 1\}$. Let $f(x) = x^t A x$ and $g(x) = x^t x - 1$. Since S^n is a compact subset and f is a continuous function, there exists $v \in S^n$, such that $f(v) = M$. Now Theorem 8.17 reveals that v is an eigenvector for A: the Lagrange criterion for a local extremum translates into the existence of $\lambda \in \mathbb{R}$, such that

$$\nabla f(v) = \lambda \nabla g(v).$$

One computes that $\nabla g(v) = 2v$. The small miracle happens in the computation of $\nabla f(v)$. Here the symmetry of the matrix A is used in a computation showing that $\nabla f(v) = 2Av$. Since $v \neq 0$, the Lagrange multiplier λ is an eigenvalue for A. $\qquad \square$

Figure 9.1: Augustin Louis Cauchy (1789–1857). French mathematician.

Theorem 9.24 is the key in the proof of Theorem 9.23. The remaining part of the proof is an induction using the orthonormalization procedure of Gram and Schmidt, which we will not enter into here. In Exercise 9.18 a (formal) proof of Theorem 9.24 is outlined avoiding the use of Theorem 8.17, but still using that a continuous function on a compact subset attains its maximum.

Example 9.25. In the simple example

$$A = \begin{pmatrix} 1 & \frac{1}{2} \\ \frac{1}{2} & 1 \end{pmatrix}$$

we find the eigenvalues λ_1 and λ_2 by solving the quadratic equation coming from the characteristic polynomial

$$\begin{aligned} \det(A - \lambda I) &= (1 - \lambda)(1 - \lambda) - (\tfrac{1}{2})^2 \\ &= \lambda^2 - 2\lambda - \tfrac{3}{4} = 0 \end{aligned}$$

giving $\lambda_1 = \frac{3}{2}$ and $\lambda_2 = \frac{1}{2}$. For $\lambda = \lambda_1, \lambda_2$ we find non-zero vectors v with $(A - \lambda I)v = 0$ and make them into an orthonormal basis. As an example consider $\lambda = \lambda_2 = \frac{1}{2}$. Here we need to find $v \neq 0$, such that

$$\begin{pmatrix} \frac{1}{2} & \frac{1}{2} \\ \frac{1}{2} & \frac{1}{2} \end{pmatrix} v = 0.$$

This is accomplished with $v = (-1, 1)$, but v is not a unit vector. We divide it by $\sqrt{v^t v} = \sqrt{2}$. The columns in the matrix

$$T = \begin{pmatrix} \frac{1}{\sqrt{2}} & -\frac{1}{\sqrt{2}} \\ \frac{1}{\sqrt{2}} & \frac{1}{\sqrt{2}} \end{pmatrix}$$

contain the coordinates of these orthonormal basis vectors. You can check that $T^{-1} = T^t$ simply by verifying that $T^t T = I$. Furthermore

$$T^t A T = \begin{pmatrix} \frac{3}{2} & 0 \\ 0 & \frac{1}{2} \end{pmatrix}$$

giving an example of Theorem 9.23 (and Theorem 9.17). It is important to notice that the reduction with

$$B = \begin{pmatrix} 1 & -\frac{1}{2} \\ 0 & 1 \end{pmatrix}$$

coming from Example 9.7 gives

$$B^t A B = \begin{pmatrix} 1 & 0 \\ 0 & \frac{3}{4} \end{pmatrix}.$$

Of course, here $B^t \neq B^{-1}$.

9.8   Quadratic forms

The function $f : \mathbb{R}^2 \to \mathbb{R}$ given by

$$f(x, y) = x^2 + 4xy + y^2 \tag{9.13}$$

is an example of a *quadratic form in the two variables* x and y. As an introductory challenge you should try, only using (9.13), to plot the points

$(x, y) \in \mathbb{R}^2$ with $f(x, y) = 1$. Quadratic forms can be simplified using reduction of real symmetric matrices. You can check that

$$f(x, y) = (x, \ y) \begin{pmatrix} 1 & 2 \\ 2 & 1 \end{pmatrix} \begin{pmatrix} x \\ y \end{pmatrix}.$$

But we already know that

$$\begin{pmatrix} 1 & 2 \\ 2 & 1 \end{pmatrix} = \begin{pmatrix} 1 & 0 \\ 2 & 1 \end{pmatrix} \begin{pmatrix} 1 & 0 \\ 0 & -3 \end{pmatrix} \begin{pmatrix} 1 & 2 \\ 0 & 1 \end{pmatrix}$$

by the example following Lemma 9.6. Therefore

$$f(x, y) = (x + 2y, \ y) \begin{pmatrix} 1 & 0 \\ 0 & -3 \end{pmatrix} \begin{pmatrix} x + 2y \\ y \end{pmatrix} = (x + 2y)^2 - 3y^2.$$

Now you can verify that

$$f(x, y) = 1 \quad \text{if and only if} \quad (x, y) = (\pm\sqrt{1 + 3t^2} - 2t, t) \qquad (9.14)$$

for $t \in \mathbb{R}$. With (9.14), plotting the points with $f(x, y) = 1$ is no more difficult than plotting the graph $(x, g(x))$ of a one variable function $g : \mathbb{R} \to \mathbb{R}$. In our example the graph is a hyperbola. If the diagonal entries from our transformation turned out non-negative it would be an ellipse. In this section we will, however, focus on the algebra of quadratic forms.

The above reduction procedure is a special case of (9.11). You start out with a quadratic form containing "mixed" terms like $4xy$ and rewrite it as a linear combination of squares of linear forms. In studying the formula for the solutions of the quadratic equation you have almost certainly seen it under the colloquial term "completing the square". In the example below, this connection to symmetric matrices is illustrated.

Example 9.26. The quadratic form

$$ax^2 + bxy + cy^2$$

can be written as

$$(x, \ y)A \begin{pmatrix} x \\ y \end{pmatrix} \qquad (9.15)$$

with

$$A = \begin{pmatrix} a & \frac{b}{2} \\ \frac{b}{2} & c \end{pmatrix}.$$

If $a \neq 0$, we put

$$B = \begin{pmatrix} 1 & -\frac{b}{2a} \\ 0 & 1 \end{pmatrix}$$

obtaining

$$B^t AB = \begin{pmatrix} a & 0 \\ 0 & -\frac{b^2}{4a} + c \end{pmatrix}$$

by Example 9.7. Therefore

$$A = \begin{pmatrix} 1 & 0 \\ \frac{b}{2a} & 1 \end{pmatrix} \begin{pmatrix} a & 0 \\ 0 & -\frac{b^2}{4a} + c \end{pmatrix} \begin{pmatrix} 1 & \frac{b}{2a} \\ 0 & 1 \end{pmatrix}.$$

Inserting this into (9.15) we get

$$ax^2 + bxy + cy^2 = a \left(x + \frac{b}{2a} y \right)^2 + \left(c - \frac{b^2}{4a} \right) y^2.$$

Putting $y = 1$ gives

$$ax^2 + bx + c = a \left(x + \frac{b}{2a} \right)^2 + \left(c - \frac{b^2}{4a} \right)$$

which is the classical completion of the square. For $ax^2 + bx + c = 0$, this gives the well known formula

$$x = \frac{-b \pm \sqrt{b^2 - 4ac}}{2a}$$

provided $b^2 - 4ac \geq 0$.

Quadratic forms in several variables

A *quadratic form in n variables* is a function $f : \mathbb{R}^n \to \mathbb{R}$ given by a sum

$$f(x_1, \ldots, x_n) = \sum_{1 \leq i,j \leq n} a_{ij} x_i x_j$$

of n^2 terms, where $a_{ij} \in \mathbb{R}$. In (9.13) $n = 2$ and putting $x_1 = x$, $x_2 = y$ we may take $a_{11} = 1$, $a_{12} = 4$, $a_{21} = 0$ and $a_{22} = 1$. As you will see, the choice of the numbers a_{ij} is in no way unique. The connection to (symmetric) matrices comes through the rewriting

$$f(x_1,\ldots,x_n) = x_1(a_{11}x_1 + \cdots + a_{1n}x_n) + \cdots + x_n(a_{n1}x_1 + \cdots + a_{nn}x_n)$$

$$= (x_1,\ldots,x_n) \begin{pmatrix} a_{11} & \cdots & a_{1n} \\ \vdots & \ddots & \vdots \\ a_{n1} & \cdots & a_{nn} \end{pmatrix} \begin{pmatrix} x_1 \\ \vdots \\ x_n \end{pmatrix}.$$

For $i \neq j$ we have $x_i x_j = x_j x_i$ and

$$a_{ij}x_i x_j + a_{ji}x_j x_i = (a_{ij} + a_{ji})/2\, x_i x_j + (a_{ij} + a_{ji})/2\, x_j x_i.$$

So replacing A by the symmetric matrix (C_{ij}) given by $C_{ij} = (a_{ij} + a_{ji})/2$ (or $C = \frac{1}{2}(A + A^t)$) we get

$$f(x) = x^t C x$$

with $x = (x_1,\ldots,x_n)^t \in \mathbb{R}^n$. We have proved that a quadratic form $f : \mathbb{R}^n \to \mathbb{R}$ can be expressed as $f(x) = x^t C x$, where C is a symmetric matrix. To fully digest these formal manipulations take a look at the following examples.

Example 9.27. If

$$f(x, y) = x^2 + 6xy + y^2 = x^2 + 3xy + 3yx + y^2,$$

then

$$f(x, y) = (x,\ y) \begin{pmatrix} 1 & 3 \\ 3 & 1 \end{pmatrix} \begin{pmatrix} x \\ y \end{pmatrix}.$$

If

$$f(x, y, z) = x^2 + 3y^2 + 5z^2 + 4xz + 10xy$$
$$= x^2 + 3y^2 + 5z^2 + 2xz + 2zx + 5xy + 5yx,$$

then

$$f(x, y, z) = (x,\ y,\ z) \begin{pmatrix} 1 & 5 & 2 \\ 5 & 3 & 0 \\ 2 & 0 & 5 \end{pmatrix} \begin{pmatrix} x \\ y \\ z \end{pmatrix}.$$

Algebraic reductions of quadratic forms

Consider the second quadratic form

$$f(x, y, z) = x^2 + 3y^2 + 5z^2 + 4xz + 10xy \qquad (9.16)$$

from Example 9.27. How would you decide if $f(x, y, z) \geq 0$ for every $x, y, z \in \mathbb{R}$ not knowing the identity (9.11)? This is not entirely obvious from scratch. We will at first focus on reducing (9.16) using algebraic manipulations. Later you will see (again?) how the algebraic reductions are mirrored in reductions of the symmetric matrix representing (9.16) using the identity (9.11). I am spelling this out, because I find it fascinating that these rather involved algebraic operations have such a nice computational framework using matrices.

The first step is to subtract a multiple of the square of a linear form $\lambda_1(ax + by + cz)^2$ from f such that we cancel all terms involving x ending up with a quadratic form in only y and z. The next step is subtracting a multiple $\lambda_2(dy + ez)^2$ ending with a quadratic form in only z. Here are these steps written out:

$$x^2 + 3y^2 + 5z^2 + 4xz + 10xy - (x + 5y + 2z)^2$$
$$= -22y^2 - 20yz + z^2 \qquad (9.17)$$
$$- 22y^2 - 20yz + z^2 + \tfrac{1}{22}(-22y - 10z)^2$$
$$= \tfrac{61}{11}\overset{*}{z^2}.$$

This shows the non-obvious algebraic identity

$$f(x, y, z) = x^2 + 3y^2 + 5z^2 + 4xz + 10xy$$
$$= (x + 5y + 2z)^2 - \tfrac{1}{22}(22y + 10z)^2 + \tfrac{61}{11}z^2. \qquad (9.18)$$

From this identity you can explicitly see that $f(-5, 1, 0) < 0$. This was certainly not obvious from the expression of f in (9.16).

Referring back to Examples 9.19 and 9.27 we get

$$f(x, y, z) = (x, \ y, \ z) \begin{pmatrix} 1 & 5 & 2 \\ 5 & 3 & 0 \\ 2 & 0 & 5 \end{pmatrix} \begin{pmatrix} x \\ y \\ z \end{pmatrix}$$

$$= (x, \ y, \ z) \begin{pmatrix} 1 & 0 & 0 \\ 5 & 1 & 0 \\ 2 & \frac{5}{11} & 1 \end{pmatrix} \begin{pmatrix} 1 & 0 & 0 \\ 0 & -22 & 0 \\ 0 & 0 & \frac{61}{11} \end{pmatrix} \begin{pmatrix} 1 & 5 & 2 \\ 0 & 1 & \frac{5}{11} \\ 0 & 0 & 1 \end{pmatrix} \begin{pmatrix} x \\ y \\ z \end{pmatrix},$$

which is in fact a true replicate of (9.18).

Positive definite and semidefinite quadratic forms

In complete analogy with the definitions for symmetric matrices we introduce the following. Let $f : \mathbb{R}^n \to \mathbb{R}$ be a quadratic form. Then f is called *positive semidefinite* if

$$f(x) \geq 0$$

for every $x \in \mathbb{R}^n$ and *positive definite* if

$$f(x) > 0$$

for every $x \in \mathbb{R}^n \setminus \{0\}$. We summarize our results in the following, where we have also added the invariance of the so-called index of the quadratic form (Sylvester's law of inertia).

Theorem 9.28. *Let $f : \mathbb{R}^n \to \mathbb{R}$ be a quadratic form given by*

$$f(x) = x^t C x,$$

where C is a real symmetric matrix. Then

(1) *f is convex if and only if C is positive semidefinite and f is strictly convex if and only if C is positive definite.*

(2) *There exists a basis $b_1, \ldots, b_n$ of $\mathbb{R}^n$ and numbers $\lambda_1, \ldots, \lambda_n$, such that*

$$f(x) = \lambda_1 (b_1^t x)^2 + \cdots + \lambda_n (b_n^t x)^2. \tag{9.19}$$

(3) *The numbers $p = |\{i \,|\, \lambda_i > 0\}|$ and $q = |\{i \,|\, \lambda_i < 0\}|$ of positive and negative signs among $\{\lambda_1, \ldots, \lambda_n\}$ are independent of the chosen expression in* (9.19).

Proof. The first statement (1), follows by definition and Theorem 9.5 using that $\nabla^2 f(x) = 2C$ for $x \in \mathbb{R}^n$ except for the claim that C is positive definite if f is strictly convex. However, if C is positive semidefinite, but not positive definite, there exists a non-zero vector $u \in \mathbb{R}^n$ with $f(u) = 0$. In this case, $0 = \lambda^2 f(u) = f(\lambda u) = f((1-\lambda) \cdot 0 + \lambda u)$ and $(1-\lambda)f(0) + \lambda f(u) = 0$ for $\lambda \in \mathbb{R}$, showing that f is not strictly convex.

The proof of (2) follows from Corollary 9.18: suppose that $C = B^t D B$, where B is invertible with row vectors $b_1, \ldots, b_n$ and D a diagonal matrix with diagonal entries $\lambda_1, \ldots, \lambda_n$. Then

$$f(x) = x^t C x = x^t B^t D B x = (Bx)^t D(Bx)$$
$$= \lambda_1 (b_1^t x)^2 + \cdots + \lambda_n (b_n^t x)^2.$$

The claim (3) was originally stated by Sylvester.[2] It has been said that he viewed the result so obvious that he did not bother to write down a proof. We will give a proof here. Suppose that

$$f(x) = \lambda_1(b_1^t x)^2 + \cdots + \lambda_n(b_n^t x)^2$$
$$= \mu_1(c_1^t x)^2 + \cdots + \mu_n(c_n^t x)^2,$$

where $b_1, \ldots, b_n$ and $c_1, \ldots, c_n$ are two bases of $\mathbb{R}^n$. Let $p = |\{i \mid \lambda_i > 0\}|$, $q = |\{i \mid \lambda_i < 0\}|$, $p' = |\{i \mid \mu_i > 0\}|$ and $q' = |\{i \mid \mu_i < 0\}|$. We will assume that $\lambda_1, \ldots, \lambda_p > 0, \lambda_{p+1} < 0, \ldots, \lambda_{p+q} < 0$ and $\mu_1 > 0, \ldots, \mu_{p'} > 0$, $\mu_{p'+1} < 0, \ldots, \mu_{p'+q'} < 0$. Rearranging we get

$$\lambda_1(b_1^t x)^2 + \cdots + \lambda_p(b_p^t x)^2 - \mu_{p'+1}(c_{p'+1}^t x)^2 - \cdots - \mu_{p'+q'}(c_{p'+q'}^t x)^2$$

equal to

$$\mu_1(c_1^t x)^2 + \cdots + \mu_{p'}(c_{p'}^t x)^2 - \lambda_{p+1}(b_{p+1}^t x)^2 - \cdots - \lambda_{p+q}(b_{p+q}^t x)^2.$$

This identity shows an inclusion $W_1 \subseteq W_2$ of subspaces $W_1, W_2 \subseteq \mathbb{R}^n$, where

$$W_1 = \{x \in \mathbb{R}^n \mid b_1^t x = \cdots = b_p^t x = c_{p'+1}^t x = \cdots = c_{p'+q'}^t x = 0\}$$
$$W_2 = \{x \in \mathbb{R}^n \mid c_1^t x = \cdots = c_{p'}^t x = c_{p'+1}^t x = \cdots = c_{p'+q'}^t x = 0\}.$$

But $\dim W_2 = n - (p' + q')$ since $c_1, \ldots, c_{p'+q'}$ are linearly independent and $\dim W_1 \geq n - (p + q')$ as the span of $b_1, \ldots, b_p, c_{p'+1}, \ldots, c_{p'+q'}$ can have dimension at most $p + q'$. Since $\dim W_1 \leq \dim W_2$ we have $n - (p + q') \leq n - (p' + q')$ giving $p' \leq p$. By symmetry we have $p \leq p'$ and therefore $p = p'$. The same reasoning shows that $q = q'$.   □

9.9   Exercises

Exercise 9.1. Show in detail that $\partial f(0) = [-1, 1]$ for $f(x) = |x|$.

Exercise 9.2. Let S be a subset and $f : S \to \mathbb{R}$. Suppose that $0 \in \partial f(x_0)$. What can you say about x_0?

Exercise 9.3. Let U be a convex open subset of $\mathbb{R}^n$ and $f : U \to \mathbb{R}$ a bounded convex function (bounded means that there exists $M \in \mathbb{R}$ such

[2]James Joseph Sylvester (1814–1897). English mathematician.

that $|f(x)| \leq M$ for every $x \in U$). Prove that for every $x_0 \in U$ there exists $\epsilon > 0$ such that

$$S(\epsilon) = \{\xi \in \partial f(x) \,|\, |x - x_0| < \epsilon\} \subseteq \mathbb{R}^n$$

is a bounded subset.

Exercise 9.4. This exercise describes the computation of $f_d''(t)$ using the chain rule noticing that $f_d'(t)$ is expressed in (9.6) as the composition of three differentiable maps:

$$g_1(t) = x_0 + td$$
$$g_2(u) = \nabla f(u)$$
$$g_3(u) = u^t d,$$

where $g_1 : \mathbb{R} \to \mathbb{R}^n$, $g_2 : \mathbb{R}^n \to \mathbb{R}^n$ and $g_3 : \mathbb{R}^n \to \mathbb{R}$. Show that $f_d'(t) = g_3 \circ g_2 \circ g_1$, where $f_d'(t)$ is given in (9.6). Show that $g_1'(t) = d$, $g_3'(u) = d^t$ and

$$g_2'(u) = \begin{pmatrix} \dfrac{\partial^2 f}{\partial x_1 \partial x_1}(u) & \cdots & \dfrac{\partial^2 f}{\partial x_1 \partial x_n}(u) \\ \vdots & \ddots & \vdots \\ \dfrac{\partial^2 f}{\partial x_n \partial x_1}(u) & \cdots & \dfrac{\partial^2 f}{\partial x_n \partial x_n}(u) \end{pmatrix}.$$

Finally prove that $f_d''(t) = d^t \nabla^2 f(x_0 + td)d$.

Exercise 9.5. Suppose that $f : U \to \mathbb{R}$ is a differentiable function, where $U \subseteq \mathbb{R}^2$ is an open subset, such that the second order partial derivatives exist at a point $v \in U$. Is it true that

$$\frac{\partial^2 f}{\partial x \, \partial y}(v) = \frac{\partial^2 f}{\partial y \, \partial x}(v)?$$

Exercise 9.6. Suppose that $f : U \to \mathbb{R}$ is a function differentiable in $x_0 \in U$, where $U \subseteq \mathbb{R}^n$ is an open subset.

(i) Prove for small $\lambda > 0$, that

$$f(x_0 + \lambda d) = f(x_0) + \lambda \nabla f(x_0)d + \lambda \epsilon(\lambda d)$$

according to Definition 8.1, where d is a unit vector.

(ii) Now suppose that $\xi \in \mathbb{R}^n$ satisfies $f(x_0 + \lambda d) \geq f(x_0) + \lambda \xi^t d$ for every unit vector d and sufficiently small $\lambda > 0$. Show that $\xi^t = \nabla f(x_0)$ using and proving the inequality

$$(\nabla f(x_0) - \xi^t)d + \epsilon(\lambda d) \geq 0$$

for $\lambda > 0$ sufficiently small. Conclude that $\partial f(x_0) = \{\nabla f(x_0)^t\}$.

Exercise 9.7. Prove that a bounded convex function $f : \mathbb{R}^n \to \mathbb{R}$ is constant.

Exercise 9.8. Prove that

$$f(x, y) = x^2 + y^2$$

is a strictly convex function from $\mathbb{R}^2$ to $\mathbb{R}$.

Exercise 9.9. Consider

$$f(x, y) = x^3 + y^3.$$

Give examples of two convex subsets C_1 and C_2 of $\mathbb{R}^2$, such that f is a convex function on C_1 but not a convex function on C_2.

Exercise 9.10. Is $f(x, y) = \cos(x) + \sin(y)$ strictly convex on some non-empty open convex subset of the plane?

Exercise 9.11. Let $f : \mathbb{R}^2 \to \mathbb{R}$ be given by

$$f(a, b) = (1 - a - b)^2 + (5 - 2a - b)^2 + (2 - 3a - b)^2.$$

(i) Show that f is a convex function. Is f strictly convex?
(ii) Find $\min\{f(a, b) \,|\, (a, b) \in \mathbb{R}^2\}$. Is this minimum unique?
(iii) Give a geometric interpretation of the minimization problem in (ii).

Exercise 9.12. Show that $f : \mathbb{R}^2 \to \mathbb{R}$ given by

$$f(x, y) = \log(e^x + e^y)$$

is a convex function. Is f strictly convex?

Exercise 9.13. Can a matrix with a 0 in the upper left hand corner be positive definite? How about positive semidefinite?

Exercise 9.14. Let us call an arbitrary $n \times n$ matrix C positive semidefinite, if $v^t C v \geq 0$ for every $v \in \mathbb{R}^n$. Suppose that A and B are positive semidefinite $n \times n$ matrices. Is the (matrix) product AB positive semidefinite? Suppose that

$$A = \begin{pmatrix} a_{11} & a_{12} \\ a_{12} & a_{22} \end{pmatrix} \quad \text{and} \quad B = \begin{pmatrix} b_{11} & b_{12} \\ b_{12} & b_{22} \end{pmatrix}$$

are symmetric and positive semidefinite. Is

$$\begin{pmatrix} a_{11} b_{11} & a_{12} b_{12} \\ a_{12} b_{12} & a_{22} b_{22} \end{pmatrix}$$

positive semidefinite? (See also Exercise 9.19.)

Exercise 9.15. Let A be a positive semidefinite $n \times n$ matrix. Suppose that $v^t A v = 0$. Prove that $Av = 0$.

Exercise 9.16. Consider the real symmetric matrix

$$A = \begin{pmatrix} a & c \\ c & b \end{pmatrix}.$$

Prove from scratch that A has a real eigenvalue (without using Theorem 9.24).

Exercise 9.17. Let

$$T = \begin{pmatrix} \cos\theta & -\sin\theta \\ \sin\theta & \cos\theta \end{pmatrix},$$

where $\theta \in \mathbb{R}$. Prove that $T^{-1} = T^t$.

Exercise 9.18. Consider the optimization problem

$$M = \max x^t A x$$
$$x \in S^n,$$

in the proof of Theorem 9.24. Suppose that $\lambda = M$ and $z^t A z = \lambda$ with $z^t z = 1$. Prove for $B = A - \lambda I$ that

(i) $z^t B z = 0$
(ii) $x^t B x \leq 0$ for every $x \in \mathbb{R}^n$.

Use this to prove $Bz = 0$ by applying that

$$(z + tBz)^t B(z + tBz) \leq 0$$

for every $t \in \mathbb{R}$. (This gives a formal alternative to the use of Lagrange multipliers in the proof of Theorem 9.24.)

Exercise 9.19. Show that any matrix of the form C^tC, where C is an $m \times n$ matrix is positive semidefinite. If $v \in \mathbb{R}^n$ show that vv^t is a positive semidefinite matrix of rank one.

Exercise 9.20. The *Hadamard product* $A \circ B$ of two matrices A and B with the same dimensions is given by $(A \circ B)_{ij} = A_{ij}B_{ij}$. Prove that $A \circ B$ is positive semidefinite if A and B are two positive semidefinite $n \times n$ matrices. Hint: use Exercise 9.19! First prove this, when B has rank one and can be written $B = vv^t$ using that

$$A \circ vv^t = D^t AD,$$

where

$$D = \begin{pmatrix} v_1 & 0 & \cdots & 0 \\ 0 & v_2 & \cdots & 0 \\ \vdots & \vdots & \ddots & \vdots \\ 0 & 0 & \cdots & v_n \end{pmatrix}$$

and $v = (v_1, \ldots, v_n)^t$. Then move on to the general case, where B is a sum of rank one positive semidefinite matrices.

Exercise 9.21. Let

$$S = \begin{pmatrix} A & 0 \\ 0 & B \end{pmatrix}$$

denote a symmetric $(r + s) \times (r + s)$ matrix, where A is a symmetric $r \times r$ matrix and B a symmetric $s \times s$ matrix. Prove that S is positive definite if and only if A and B are positive definite. Prove also the same statement with definite replaced by semidefinite.

Exercise 9.22. Give an example of a non-symmetric $n \times n$ matrix A, such that $x^t Ax \geq 0$ for every $x \in \mathbb{R}^n$.

Exercise 9.23. Compute the set

$$\left\{ (a, b) \in \mathbb{R}^2 \,\middle|\, \begin{pmatrix} 2 & 1 & a \\ 1 & 1 & 1 \\ a & 1 & b \end{pmatrix} \text{ is positive definite} \right\}.$$

Same question with positive semidefinite. Sketch and compare the two subsets of the plane.

Exercise 9.24. Let $f : \mathbb{R}^2 \to \mathbb{R}$ be given by

$$f(x, y) = ax^2 + by^2 + cxy,$$

where $a, b, c \in \mathbb{R}$.

(i) Show that f is a convex function if and only if $a \geq 0, b \geq 0$ and $4ab - c^2 \geq 0$.

(ii) Suppose now that $a > 0$ and $4ab - c^2 > 0$. Show that $g(x, y) = f(x, y) + x + y$ has a unique global minimum and give a formula for this minimum in terms of a, b and c.

In the following we will assume that f is a convex function.

(iii) Show that f is not strictly convex if $a = 0$.

(iv) Show that f is not strictly convex if $a > 0$ and $4ab - c^2 = 0$.

Exercise 9.25. Recall that a convex cone in a euclidean vector space $\mathbb{R}^n$ is a subset $C \subseteq \mathbb{R}^n$ such that $x + y \in C$ and $\lambda x \in C$ for every $x, y \in C$ and every $\lambda \geq 0$. Let

$$S = \left\{ \begin{pmatrix} a & b & c \\ b & d & e \\ c & e & f \end{pmatrix} \,\middle|\, a, b, c, d, e, f \in \mathbb{R} \right\}$$

be the set of symmetric 3×3 matrices. Notice that S may be viewed as the vector space $\mathbb{R}^6 = \{(a, b, c, d, e, f) \,|\, a, b, c, d, e, f \in \mathbb{R}\}$.

(i) Let $x \in \mathbb{R}^3$. Show that

$$C_x = \{A \in S \,|\, x^t A x \geq 0\}$$

is a closed convex cone S.

(ii) Let $\mathcal{P} \subseteq S$ be the set of positive semidefinite matrices. Show that

$$\mathcal{P} = \bigcap_{x \in \mathbb{R}^n} C_x.$$

Use this to prove that $\mathcal{P}$ is a closed convex cone in S.

(iii) Prove that the closure of the set of positive definite matrices inside S coincides with $\mathcal{P}$.

(iv) Let

$$S' = \left\{ \begin{pmatrix} a & b & c \\ b & d & e \\ c & e & f \end{pmatrix} \in \mathcal{P} \,\middle|\, a - b + c = 0, \ e = 0 \right\}.$$

Show that S' is a convex set in S.

Exercise 9.26. Why is the subset given by the inequalities

$$x \geq 0$$
$$y \geq 0$$
$$xy - z^2 \geq 0$$

a convex subset of $\mathbb{R}^3$?

Exercise 9.27. Let S denote the convex cone of positive semidefinite $n \times n$ matrices in the vector space of symmetric $n \times n$ matrices. Prove that

$$\{vv^t \,|\, v \in \mathbb{R}^n \setminus \{0\}\}$$

are the extreme rays in S (use Exercise 9.15 and Corollary 9.18).

Exercise 9.28. Let

$$f(x, y, z) = 3x^2 - 2xy - 2xz + 3y^2 + 3z^2 - 2yz.$$

(i) Show that $f(x, y, z) \geq 0$ for every $x, y, z \in \mathbb{R}$.
(ii) Show that $f(x, y, z) = 0$ if and only if $x = 0$, $y = 0$ and $z = 0$.

Exercise 9.29. Let $f_{(a,b)}$ be given by

$$f_{(a,b)}(x, y, z) = ax^2 + by^2 + 4z^2 + 2xy + 4xz + 6yz$$

and let S denote the set of (a, b) such that $f_{(a,b)}$ is strictly convex. Show that S is non-empty and convex. Give a finite set of inequalities that defines S.

Exercise 9.30. Let

$$f(x, y) = 5x^2 - 2xy + 5y^2.$$

Based on Example 9.25, prove that $f(x, y) \geq 0$ using an orthonormal basis of eigenvectors for a symmetric matrix associated with f.

Exercise 9.31. Let

$$f(x,y) = 2x^2 + 3y^2 + 4xy.$$

Prove that $f(x,y) \geq 0$ for every $x, y \in \mathbb{R}$. Plot the points $(x,y) \in \mathbb{R}^2$ with $f(x,y) = 1$.

Exercise 9.32. Is the quadratic form

$$f(x,y,z) = 2x^2 + 2y^2 + 4z^2 + 2xy + 4xz + 6yz$$

positive semidefinite?

Exercise 9.33. Let

$$f(x,y,z) = x^2 + 3y^2 + 5z^2 + 4xz + 10xy.$$

Compute a basis $b_1, b_2, b_3 \in \mathbb{R}^3$, such that

$$f(v) = \lambda_1 (b_1^t v)^2 + \lambda_2 (b_2^t v)^2 + \lambda_3 (b_3^t v)^2$$

for $v = (x,\ y,\ z)^t$. Compute p and q as given in Theorem 9.28 for f. Find $x, y, z \in \mathbb{R}$ with $f(x,y,z) < 0$.

[Extra credit!] Use a computer to sketch the set

$$\{(x,\ y,\ z) \in \mathbb{R}^3 \,|\, f(x,y,z) = 1\}.$$

Exercise 9.34. Let $f_a : \mathbb{R}^2 \to \mathbb{R}$ denote the quadratic form given by

$$f_a(x,y,z) = x^2 + y^2 + z^2 + 2axy + 2xz + 2yz,$$

where $a \in \mathbb{R}$.

(i) Write down the symmetric matrix associated to f_a.
(ii) Show that f_a is not positive definite for $a = 2$.
(iii) Show that f_a is not positive definite for any $a \in \mathbb{R}$.
(iv) Is there an $a \in \mathbb{R}$ such that f_a is positive semidefinite?
(v) Rewrite f_a for $a = 1$ as a sum of squares of linear forms in x, y and z.

Chapter 10

Convex optimization

In [Sylvester (1857)] the following problem was posed. Given n points $(x_1, y_1), \ldots, (x_n, y_n) \in \mathbb{R}^2$, what is the center and radius of the smallest circle containing these points? A potential real world application could be placing an alarm central so that the maximal distance to a house is minimized.

At first sight, this problem (see Figure 10.1) can be converted into finding the smallest r, such that

$$(x - x_i)^2 + (y - y_i)^2 \leq r^2 \qquad \text{for } i = 1, \ldots, n$$

for some $x, y \in \mathbb{R}$. A little rewriting turns this into the problem of minimizing the convex function $f(x, y, \lambda) = x^2 + y^2 + 2\lambda$ subject to the polyhedral constraints

$$\tfrac{1}{2}(x_i^2 + y_i^2) \leq x_i x + y_i y + \lambda \qquad \text{for } i = 1, \ldots, n,$$

with $r^2 = x^2 + y^2 + 2\lambda$. This is an example of a non-linear convex optimization problem in contrast to the linear programming problems encountered earlier (see Example 1.2 and §5.4).

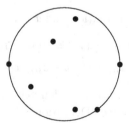

Figure 10.1: Smallest circle containing the seven marked points.

In the most general setting, an *optimization problem* in $\mathbb{R}^n$ has the form

$$f^* := \inf\{f(x) \mid x \in S\}, \tag{10.1}$$

where $S \subseteq \mathbb{R}^n$ can be any subset and $f : S \to \mathbb{R}$ any function. A point $x \in \mathbb{R}^n$ is called *feasible* for (10.1) if $x \in S$. An *optimal solution* for (10.1) is a feasible solution $x_0 \in S$ with $f(x_0) = f^*$ i.e.,

$$f(x_0) \leq f(x)$$

for every $x \in S$. In this case $f(x_0)$ is called the *optimal value* of (10.1). If $f^* = -\infty$ the optimization problem does not have an optimal solution. By abuse of notation we will sometimes write min instead of inf in (10.1).

In this chapter we introduce the famous Karush-Kuhn-Tucker conditions for optimality in (10.1), show how the saddle points from game theory lead to the concept of a dual optimization problem and outline an interior point computational procedure for solving (10.1) in the convex case.

This chapter only gives a brief introduction to the vast field of convex optimization. For further study, the book [Boyd and Vandenberghe (2004)] is highly recommended.

We begin by stating a geometric criterion for optimality in (10.1) under suitable assumptions of convexity and differentiability.

10.1   A geometric optimality criterion

Proposition 10.1. *Let $S \subseteq \mathbb{R}^n$ be a convex subset and $f : U \to \mathbb{R}$ a differentiable function defined on an open subset $U \supseteq S$. If $x_0 \in S$ is an optimal solution of (10.1), then*

$$\nabla f(x_0)(x - x_0) \geq 0 \qquad \text{for every } x \in S. \tag{10.2}$$

If f in addition is a convex function, then (10.2) implies that x_0 is optimal.

Proof.   If x_0 is an optimal solution and $x \in S \setminus \{x_0\}$, then

$$0 \leq f((1-t)x_0 + tx) - f(x_0) = f(x_0 + t(x - x_0)) - f(x_0)$$
$$= t\,(\nabla f(x_0)(x - x_0) + \epsilon(t(x - x_0))|x - x_0|)$$

for every t with $0 \leq t \leq 1$, where ϵ denotes the epsilon function in the definition of differentiability (see Definition 8.1). Therefore

$$\nabla f(x_0)(x - x_0) + \epsilon(t(x - x_0))|x - x_0| \geq 0$$

for $0 \leq t \leq 1$. This is only possible if $\nabla f(x_0)(x - x_0) \geq 0$. We have silently applied the convexity of S and the differentiability of f at x_0.

If f in addition is convex and (10.2) holds, then Theorem 9.1 shows that x_0 is an optimal solution. □

If you compare Definition 6.10 and Proposition 10.1 assuming that f is convex, you will discover a nice geometric interpretation: a point x_0 is optimal if and only if $\nabla f(x_0)x = \nabla f(x_0)x_0$ is a supporting hyperplane for S (see Figure 10.2) assuming that $\nabla f(x_0) \neq 0$.

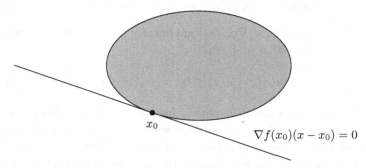

Figure 10.2: Optimality of $x_0 = (-\frac{1}{2}, -\frac{1}{2})$ in $\min\{f(x,y) \,|\, x^2 + 3y^2 \leq 1\}$ for $f(x,y) = (x+1)^2 + (y+2)^2$ proved using Proposition 10.1.

Example 10.2. Let $S = \text{conv}(\{x_1, \ldots, x_m\}) \subseteq \mathbb{R}^n$ and consider the optimization problem $M = \min\{f(v) \,|\, v \in S\}$ for $f(v) = c^t v$ and $c \in \mathbb{R}^n$. In this case $\nabla f(x_0) = c^t$ and Proposition 10.1 can be simplified to $x_0 \in S$ optimal if and only if $c^t(x_i - x_0) \geq 0$ for every $i = 1, \ldots, m$, since

$$c^t(\lambda_1 x_1 + \cdots + \lambda_m x_m - x_0) = \lambda_1 c^t(x_1 - x_0) + \cdots + \lambda_m c^t(x_m - x_0)$$

for $\lambda_i \geq 0$ and $\lambda_1 + \cdots + \lambda_m = 1$. This also shows that a minimum of f on S is attained in one of the points $x_1, \ldots, x_m$.

Example 10.3. The differentiable function $f : \mathbb{R}^2 \to \mathbb{R}$ given by

$$f(x,y) = x^2 - 4xy + y^2$$

is not convex and $\nabla f(x,y) = (2x - 4y, 2y - 4x)$. Let $S = \{(x,y) \,|\, 0 \leq x \leq 1, -1 \leq y \leq 0\}$ and $x_0 = (0,0)$. Even though $\nabla f(x_0)(x - x_0) \geq 0$ for every $x \in S$, x_0 is not optimal. Consider $x_0 = (1, -1) \in S$ with $f(x_0) = -2$ and $\nabla f(x_0) = (6, -6)$. Since $\nabla f(x_0)(x - x_0) < 0$ for $x = (1, 0)$, Proposition 10.1 says that x_0 is not optimal.

10.2   The Karush-Kuhn-Tucker conditions

In §8.3 we studied the optimization problem

$$\min\{f(x) \mid g_1(x) = 0, \ldots, g_m(x) = 0\}$$

for differentiable functions $f, g_1, \ldots, g_m : \mathbb{R}^n \to \mathbb{R}$ and proved that if x_0 is an optimal solution, there exists (Lagrange multipliers) $\lambda_1, \ldots, \lambda_m \in \mathbb{R}$, such that

$$\nabla f(x_0) + \lambda_1 \nabla g_1(x_0) + \cdots + \lambda_m \nabla g_m(x_0) = 0 \qquad (10.3)$$

provided that $\nabla g_1(x_0), \ldots, \nabla g_m(x_0)$ are linearly independent. Here we will consider the optimization problem

$$\min\{f(x) \mid g_1(x) \leq 0, \ldots, g_m(x) \leq 0\}, \qquad (10.4)$$

where the constraints are inequalities. Inequalities and gradients for differentiable functions interact beautifully with polyhedral convexity in giving new and interesting conditions for optimality that adds inequalities to (10.3). These conditions date back to the master's thesis [Karush (1939)] by Karush[1] and the seminal paper [Kuhn and Tucker (1951)] by Kuhn[2] and Tucker[3] and are as follows.

Definition 10.4. The system

$$\lambda_i \geq 0$$
$$g_i(x_0) \leq 0 \quad \text{and} \quad \lambda_i g_i(x_0) = 0$$
$$\nabla f(x_0) + \lambda_1 \nabla g_1(x_0) + \cdots + \lambda_m \nabla g_m(x_0) = 0,$$

of inequalities for $i = 1, \ldots, m$ in the unknowns $x_0 \in \mathbb{R}^n$ and $\lambda_1, \ldots, \lambda_m \in \mathbb{R}$ are called the *Karush-Kuhn-Tucker (KKT) conditions* associated with the optimization problem (10.4).

The classical Lagrange multiplier conditions in Remark 8.18 form a system of $m + n$ equations with $m + n$ unknowns. In Definition 10.4 there are $2m$ inequalities and $m+n$ equations in $m+n$ unknowns. Notice in particular the "complementary" condition $\lambda_i g_i(x_0) = 0$. If $\lambda_i > 0$, this condition dictates that $g_i(x_0) = 0$. Similarly if $g_i(x_0) < 0$ we must have $\lambda_i = 0$.

---

[1] William Karush (1917–1997). American mathematician.
[2] Harold Kuhn (1925–). American mathematician.
[3] Albert Tucker (1905–1995). American mathematician.

For reasons that will become clear soon, we need to introduce the following important concept.

Definition 10.5. The optimization problem (10.4) is called *strictly feasible* if there exists $z_0 \in \mathbb{R}^n$ with

$$g_1(z_0) < 0$$
$$\vdots$$
$$g_m(z_0) < 0.$$

Historically this condition seems to originate in a Cowles Commission report by Morton Slater [Slater (1950)] and it is also referred to as the *Slater condition*. It turns out to be quite important in duality theory for optimization (see Theorem 10.9). If you study the proof of Theorem 10.6(2), I am sure you will agree that strong feasibility as a replacement for linear independence of the gradients is quite clever.

Below we will show how the KKT conditions appear naturally as a necessary condition for optimality using only polyhedral convexity: Gordan's theorem applies when the gradients are linearly independent. Farkas's lemma applies under the assumption of convex constraints and strict feasibility. Notice in the proof of Theorem 10.6(1) how Gordan's theorem (Theorem 4.17) almost instantly gives the Karush-Kuhn-Tucker conditions.

Theorem 10.6. *Consider the optimization problem*

$$\min\{f(x) \mid x \in S\} \tag{10.5}$$

where $S = \{x \in \mathbb{R}^n \mid g_1(x) \leq 0, \ldots, g_m(x) \leq 0\}$ with $g_1, \ldots, g_m : \mathbb{R}^n \to \mathbb{R}$ differentiable functions and $f : U \to \mathbb{R}$ differentiable on an open set U containing S. Let $x_0 \in S$ denote an optimal solution of (10.5).

(1) *If*
$$\nabla g_1(x_0), \ldots, \nabla g_m(x_0)$$
are linearly independent, then the KKT conditions are satisfied at x_0 for suitable $\lambda_1, \ldots, \lambda_m$.

(2) *If $g_1, \ldots, g_m$ are convex and (10.5) is strictly feasible, then the KKT conditions are satisfied at x_0 for suitable $\lambda_1, \ldots, \lambda_m$.*

(3) *If $g_1, \ldots, g_m$ are affine functions i.e., $S = \{x \in \mathbb{R}^n \mid Ax \leq b\}$ for an $m \times n$ matrix A and $b \in \mathbb{R}^m$, then the KKT conditions are satisfied at x_0 for suitable $\lambda_1, \ldots, \lambda_m$.*

(4) *If f and $g_1, \ldots, g_m$ are all convex functions and the KKT conditions hold at z for some $\lambda_1, \ldots, \lambda_m \in \mathbb{R}$, then z is an optimal solution of (10.5).*

Proof.   Before writing out the proofs we need a general remark. If x_0 is a local extremum satisfying $g_1(x_0) < 0, \ldots, g_m(x_0) < 0$, then $\nabla f(x_0) = 0$ by Theorem 8.10. In this case, the KKT conditions are always satisfied with $\lambda_1 = \cdots = \lambda_m = 0$.

Proof of part (1).   Let $x_0 \in S$ be an optimal solution. Suppose that $g_1(x_0) = \cdots = g_r(x_0) = 0$ and $g_{r+1}(x_0) < 0, \ldots, g_m(x_0) < 0$ with $r \geq 1$. If $d \in \mathbb{R}^n$ satisfies

$$\nabla f(x_0)d < 0, \quad \nabla g_1(x_0)d < 0, \quad \cdots \quad , \nabla g_r(x_0)d < 0, \qquad (10.6)$$

then we can choose $\epsilon > 0$ small enough such that

$$g_j(x_0 + \lambda d) < 0$$

for every $0 \leq \lambda < \epsilon$ and $j = r+1, \ldots, m$. Now Lemma 8.9 applied to $u = d$, x_0 and $f, g_1, \ldots, g_r$ shows that x_0 cannot be optimal. Therefore the system (10.6) of inequalities does not have a solution and Theorem 4.17 ensures the existence of $\lambda_0, \lambda_1, \ldots, \lambda_r \geq 0$ not all zero, such that

$$\lambda_0 \nabla f(x_0) + \lambda_1 \nabla g_1(x_0) + \cdots + \lambda_r g_r(x_0) = 0.$$

Since $\nabla g_1(x_0), \ldots, \nabla g_m(x_0)$ are assumed linearly independent, we can take $\lambda_0 = 1$. This makes the KKT conditions satisfied with $\lambda_1, \ldots, \lambda_r$ and $\lambda_{r+1} = \cdots = \lambda_m = 0$.

Proof of part (2).   Suppose that (10.4) is strictly feasible by $z_0 \in \mathbb{R}^n$ in Definition 10.5 and that x_0 is an optimal point. By Theorem 9.1,

$$0 > g_i(z_0) \geq g_i(x_0) + \nabla g_i(x_0)(z_0 - x_0)$$

for $i = 1, \ldots, m$. With the assumption $g_1(x_0) = \cdots = g_r(x_0) = 0$ and $g_{r+1}(x_0) < 0, \ldots, g_m(x_0) < 0$ we therefore get $\nabla g_i(x_0)(z_0 - x_0) < 0$ for $i = 1, \ldots, r$. If there does not exist $\lambda_1, \ldots, \lambda_r \geq 0$ with

$$\nabla f(x_0) + \lambda_1 \nabla g_1(x_0) + \cdots + \lambda_r \nabla g_r(x_0) = 0,$$

there exists $u \in \mathbb{R}^n$, such that

$$\nabla f(x_0)u < 0$$
$$\nabla g_1(x_0)u \leq 0$$
$$\vdots$$
$$\nabla g_r(x_0)u \leq 0$$

by Lemma 4.14 (Farkas's lemma). Walking from x_0 in the direction of $z_0 - x_0$ makes the g_i smaller, since $\nabla g_i(x_0)(z_0 - x_0) < 0$ for $i = 1, \ldots, r$. Walking from x_0 in the direction of u makes f smaller, since $\nabla f(x_0)u < 0$ and does not increase g_i as $\nabla g_i(x_0)u \leq 0$ for $i = 1, \ldots, r$. We combine u and $z_0 - x_0$ in using the direction vector

$$d_\epsilon = (1 - \epsilon)u + \epsilon(z_0 - x_0)$$

for suitable $\epsilon > 0$. We have $\nabla g_j(x_0)d_\epsilon < 0$ for $j = 1, \ldots, r$ and $0 < \epsilon \leq 1$. We may pick $\epsilon > 0$ sufficiently small so that $\nabla f(x_0)d_\epsilon < 0$. Now choose $\lambda > 0$ small enough so that $g_j(x_0 + \lambda d_\epsilon) \leq 0$ for $j = r + 1, \ldots, m$. Then $x_0 + \lambda d_\epsilon \in S$ with $f(x_0 + \lambda d_\epsilon) < f(x_0)$ proving that x_0 cannot be optimal.

***Proof of part* (3).** We retrace the proof of (2) to see that strict feasibility is not needed in this case. Assuming that $g_i(x) = \alpha^t x + \beta$ for $\alpha \in \mathbb{R}^n$ and $\beta \in \mathbb{R}$, we must have

$$g_i(x) = \nabla g_i(x_0)x + \beta. \tag{10.7}$$

Again by Farkas's lemma, if there does not exist $\lambda_1, \ldots, \lambda_r \geq 0$ with

$$\nabla f(x_0) + \lambda_1 \nabla g_1(x_0) + \cdots + \lambda_r \nabla g_r(x_0) = 0,$$

there exists $u \in \mathbb{R}^n$ with

$$\nabla f(x_0)u < 0$$
$$\nabla g_1(x_0)u \leq 0$$
$$\vdots$$
$$\nabla g_r(x_0)u \leq 0.$$

If $\nabla g_i(x_0)u = 0$, then $g_i(x_0 + tu) = g_i(x_0)$ for every $t \in \mathbb{R}$ by (10.7). Therefore $g_i(x_0 + tu) \leq 0$ for small $t > 0$ and every $i = 1, \ldots, m$ contradicting that x_0 is optimal, since $f(x_0 + tu) < f(x_0)$.

Proof of part (4).   Suppose that $x_0 \in S$ satisfies the KKT conditions with $\lambda_1, \ldots, \lambda_m \geq 0$. If $\lambda_1 = \cdots = \lambda_m = 0$, then $\nabla f(x_0) = 0$ and x_0 is a global minimum according to Corollary 9.2.

Assume that $\lambda_1, \ldots, \lambda_r > 0$ and $\lambda_{r+1} = \cdots = \lambda_m = 0$ for $r \geq 1$. Hence $g_1(x_0) = \cdots = g_r(x_0) = 0$ and the "gradient" equation reads

$$\nabla f(x_0) + \lambda_1 \nabla g_1(x_0) + \cdots + \lambda_r \nabla g_r(x_0) = 0. \tag{10.8}$$

If x_0 is not a global minimum, there exists $x \in S$ with $f(x) < f(x_0)$. Since

$$f(x) \geq f(x_0) + \nabla f(x_0)(x - x_0)$$

by Theorem 9.1, we have $\nabla f(x_0)(x - x_0) < 0$. By convexity of the functions g_i, S is a convex set. Therefore $(1 - t)x_0 + tx = x_0 + t(x - x_0) \in S$ and $g_i(x_0 + t(x - x_0)) \leq 0$ for every $0 \leq t \leq 1$. This implies for $i = 1, \ldots, r$, that $\nabla g_i(x_0)(x - x_0) \leq 0$ by Lemma 8.9: if $\nabla g_i(x_0)(x - x_0) > 0$ for some i, then $g_i(x_0 + t(x - x_0)) > 0$ for small $t > 0$ (we have assumed $g_1(x_0) = \cdots = g_r(x_0) = 0$) . Having a vector $d = x - x_0$ with $\nabla f(x_0)d < 0$ and $\nabla g_1(x_0)d \leq 0, \ldots, \nabla g_r(x_0)d \leq 0$ is impossible if (10.8) holds. This proves that the KKT conditions at a point x_0 implies that x_0 is optimal assuming that all the functions involved are convex.   □

10.3   An example

Let C denote the set (see Figure 10.3) of points $(x, y) \in \mathbb{R}^2$ with

$$x^2 + 2y^2 \leq 1$$
$$x + y \leq 1$$
$$y \leq x.$$

We will illustrate the mechanics of solving the KKT conditions in finding an optimal solution for

$$\min\{x + 3y \mid (x, y) \in C\}. \tag{10.9}$$

Putting

$$g_1(x, y) = x^2 + 2y^2 - 1$$
$$g_2(x, y) = x + y - 1$$
$$g_3(x, y) = y - x$$

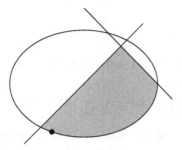

Figure 10.3: The convex set C with optimal solution for (10.9) marked.

and $f(x, y) = x + 3y$, we are in a position to apply (2) of Theorem 10.6, since g_1, g_2, g_3 are convex functions and $g_1(z_0) < 0, g_2(z_0) < 0, g_3(z_0) < 0$ for $z_0 = (0, -\frac{1}{2})$. This means that an optimal solution of (10.9) satisfies the KKT conditions. Part (4) of the same theorem tells us that the x_0 in a solution of the KKT conditions is an optimal solution (here we also use that f is a convex function). The full set of KKT conditions in $x, y, \lambda_1, \lambda_2, \lambda_3 \in \mathbb{R}$ are

$$x^2 + 2y^2 - 1 \leq 0$$
$$x + y - 1 \leq 0$$
$$y - x \leq 0$$
$$\lambda_1, \lambda_2, \lambda_3 \geq 0$$
$$\lambda_1(x^2 + 2y^2 - 1) = 0$$
$$\lambda_2(x + y - 1) = 0$$
$$\lambda_3(-x + y) = 0$$
$$1 + 2\lambda_1 x + \lambda_2 - \lambda_3 = 0$$
$$3 + 4\lambda_1 y + \lambda_2 + \lambda_3 = 0.$$

A strategy for finding a solution to the KKT conditions is trying (the eight) different combinations of strict inequalities in $\lambda_1, \lambda_2, \lambda_3 \geq 0$. You can see from the last two equations that $\lambda_1 = 0$ is impossible. The condition $\lambda_1 > 0$ shows that an optimal solution has to occur on the lower arc in Figure 10.3. If $\lambda_3 > 0$, then $x = y$ and $\lambda_2 = 1 + 3\lambda_3 > 0$ by the last two equations. This implies $x = y = \frac{1}{2}$ violating $x^2 + 2y^2 - 1 = 0$. Therefore $\lambda_3 = 0$. If $\lambda_2 > 0$, then $y = 1 - x$ and $5 + 4\lambda_1 + 3\lambda_2 = 0$ by $\lambda_3 = 0$ and the last two equations. Therefore $\lambda_2 = 0$. So we are left with the case $\lambda_1 > 0$ and $\lambda_2 = \lambda_3 = 0$

giving

$$x = -\frac{1}{2\lambda_1} \quad \text{and} \quad y = -\frac{3}{4\lambda_1}.$$

Inserting this into $x^2 + 2y^2 - 1 = 0$ we end up with (see Figure 10.3)

$$\lambda_1 = \frac{\sqrt{11}}{2\sqrt{2}}, \quad x = -\sqrt{\frac{2}{11}} \quad \text{and} \quad y = -\frac{3}{\sqrt{22}}.$$

Theorem 10.6 is beautiful mathematics. Going through the KKT conditions as above can be quite lengthy if not impossible in practice. There are more effective methods for (at least) approximating an optimal solution (see §10.5).

10.4 The Langrangian, saddle points, duality and game theory

Let $f, g_1, \ldots, g_m$ be functions $\mathbb{R}^n \to \mathbb{R}$ with no further assumptions as to differentiability, convexity etc. To the optimization problem

$$\min\{f(x) \mid x \in S\} \tag{10.10}$$

with $S = \{x \in \mathbb{R}^n \mid g_1(x) \leq 0, \ldots, g_m(x) \leq 0\}$, we associate the *Lagrangian* $L : \mathbb{R}^n \times \mathbb{R}^m \to \mathbb{R}$ given by

$$L(x, \lambda) = f(x) + \lambda_1 g_1(x) + \cdots + \lambda_m g_m(x).$$

A *saddle point* for L is a point $(x_0, \lambda^0) \in \mathbb{R}^n \times \mathbb{R}^m$ with $x_0 \in S$ and $\lambda^0 \geq 0$, such that

$$L(x_0, \lambda^0) \leq L(x, \lambda^0) \qquad \text{for every } x \in S$$

$$L(x_0, \lambda) \leq L(x_0, \lambda^0) \qquad \text{for every } \lambda \geq 0.$$

If (x_0, λ^0) is a saddle point for L, then x_0 is an optimal point for (10.10). The Lagrange multiplier λ^0 in a saddle point (x_0, λ^0) plays the role of an optimal solution in an optimization problem dual to (10.10): let $q : \mathbb{R}^m \to \mathbb{R} \cup \{-\infty\}$ be defined by

$$q(\lambda) = \inf\{L(x, \lambda) \mid x \in \mathbb{R}^n\}.$$

The optimization problem

$$\sup\{q(\lambda) \mid \lambda \geq 0\} \tag{10.11}$$

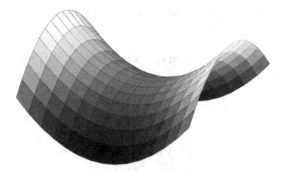

Figure 10.4: The function $f : \mathbb{R} \times \mathbb{R} \to \mathbb{R}$ given by $f(x,y) = x^2 - y^2$ is depicted above for $x, y \in [-4, 4]$. Here $(0, 0)$ is a saddle point for f, since $f(0, 0) \leq f(x, 0)$ for every $x \in \mathbb{R}$ and $f(0, y) \leq f(0, 0)$ for every $y \in \mathbb{R}$.

is called the (Lagrange) *dual* to (10.10). It is not too hard to verify that λ^0 is an optimal solution to (10.11) if (x_0, λ^0) is a saddle point for the Lagrangian. Since, $L(x, \lambda) \leq f(x)$ for $x \in S$ and $\lambda \geq 0$, the optimal value for the dual problem

$$q^* \leq f^* \qquad (10.12)$$

is a lower bound for the optimal value of the original problem.

Remark 10.7. The notion of a saddle point arises in the context of two person games (see [von Neumann and Morgenstern (1944)]). Here a game is abstractly defined as a (payoff) function $f : S \times T \to \mathbb{R}$, where S is the set of strategies of the first person and T those of the second person. A round of the game consists of the first and the second person independently picking $s \in S$ and $t \in T$ respectively. The interpretation is that the first player pays $f(s, t)$ to the second player. The first person wants to choose s minimizing

$$\sup\{f(s, t) \,|\, t \in T\},$$

whereas the second person wants to pick t maximizing

$$\inf\{f(s, t) \,|\, s \in S\}.$$

A saddle point (s_0, t^0) for f can be interpreted as an equilibrium of the game in the sense that s_0 and t^0 are optimal strategies for the first and second person. The first person is guaranteed a loss of at most $f(s_0, t^0)$ and the second person is guaranteed a gain of at least the same amount. If the first person deviates from s_0 picking s, then he may have to pay more if

$f(s_0, t^0) < f(s, t^0)$. Similarly the second person may end up gaining less with $t \in T$ if $f(s_0, t) < f(s_0, t^0)$.

Example 10.8. Consider the linear program $\min\{c^t x \mid Ax \leq b\}$. For this optimization problem,

$$L(x, \lambda) = c^t x + \lambda^t (Ax - b) = (c^t + \lambda^t A)x - \lambda^t b.$$

Therefore

$$q(\lambda) = \begin{cases} -\lambda^t b & \text{if } c^t + \lambda^t A = 0 \\ -\infty & \text{otherwise} \end{cases}$$

and the dual problem becomes

$$\max\{-b^t \lambda \mid A^t \lambda = -c, \, \lambda \geq 0\}$$

with the convention that $\max \emptyset = -\infty$ (if the dual problem is not feasible i.e., there exists no $\lambda \geq 0$ such that $A^t \lambda = -c$, then $q(\lambda) = -\infty$ for every $\lambda \geq 0$).

For the linear program $\min\{c^t x \mid Ax = b, x \geq 0\}$ in standard form, the Lagrangian is

$$\begin{aligned} L(x, \lambda_1, \lambda_2, \mu) &= c^t x + \lambda_1^t (Ax - b) + \lambda_2^t (b - Ax) - \mu^t x \\ &= (c^t - (\lambda_2 - \lambda_1)^t A - \mu^t)x + (\lambda_2 - \lambda_1)^t b, \end{aligned}$$

where $\lambda_1, \lambda_2 \in \mathbb{R}^m$ and $\mu \in \mathbb{R}^n$ correspond to the constraints $Ax - b \leq 0$, $b - Ax \leq 0$ and $-x \leq 0$ respectively. Therefore

$$q(\lambda_1, \lambda_2, \mu) = \begin{cases} (\lambda_1 - \lambda_2)^t b & \text{if } c^t - (\lambda_1 - \lambda_2)^t A = \mu^t \\ -\infty & \text{otherwise}. \end{cases}$$

Introducing $y := \lambda_1 - \lambda_2$, the dual problem becomes

$$\max\{y^t b \mid y^t A \leq c^t\}. \tag{10.13}$$

This is the dual linear program encountered in (4.16).

Equality in (10.12) is a deeper statement needing further conditions.

Theorem 10.9. *Let $f, g_1, \ldots, g_m : \mathbb{R}^n \to \mathbb{R}$ be convex differentiable functions. If x_0 is an optimal solution of*

$$\min\{f(x) \mid g_1(x) \leq 0, \ldots, g_m(x) \leq 0\} \tag{10.14}$$

and (10.14) *is strictly feasible or* $g_1, \ldots, g_m$ *are affine functions, then there exists an optimal solution* λ^0 *of the dual problem* $\max\{q(\lambda) \mid \lambda \geq 0\}$ *associated to* (10.14) *with optimal value* $q^* = f^*$.

Proof. By Theorem 10.6 there exists a solution (x_0, λ^0) to the KKT conditions for (10.14). By assumption

$$h(x) := L(x, \lambda^0)$$

is a convex differentiable function $\mathbb{R}^n \to \mathbb{R}$. By the KKT conditions, $\nabla h(x_0) = 0$ and therefore $L(x_0, \lambda^0) \leq L(x, \lambda^0)$ for every feasible x. This implies

$$q(\lambda^0) = L(x_0, \lambda^0) = f(x_0),$$

where the last $=$ follows by the KKT conditions. Therefore $q^* = f^*$. $\square$

The dual of (10.13) in Example 10.8 is the linear program $\min\{c^t x \mid Ax = b, x \geq 0\}$ in standard form. This shows that Theorem 4.16 (duality for linear programming) is a special case of Theorem 10.9.

Example 10.10. Consider the optimization problem

$$\min\{(x+1)^2 + (y+2)^2 \mid x^2 + 3y^2 \leq 1\} \tag{10.15}$$

from Figure 10.2. This problem is strictly feasible, the feasible set is compact and the functions differentiable and convex. Therefore Theorem 10.9 applies. The Lagrangian is

$$\begin{aligned}
L(x, y, \lambda) &= (x+1)^2 + (y+2)^2 + \lambda(x^2 + 3y^2 - 1) \\
&= (1+\lambda)x^2 + (1+3\lambda)y^2 + 2x + 4y + 5 - \lambda.
\end{aligned}$$

For fixed $\lambda \geq 0$ we compute the gradient of the convex function $L(x, y, \lambda)$ and find that minimum occurs for

$$x = -\frac{1}{1+\lambda} \quad \text{and} \quad y = -\frac{2}{1+3\lambda}. \tag{10.16}$$

Therefore

$$q(\lambda) = 5 - \lambda - \frac{1}{1+\lambda} - \frac{4}{1+3\lambda}. \tag{10.17}$$

By duality we have turned the two variable optimization problem (10.15) into the classical problem of finding maxima for the one variable function in (10.17).

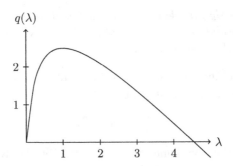

Figure 10.5: Graph of "dual" function in (10.17) associated with the optimization problem (10.15) for $0 \leq \lambda \leq 5$. Maximum occurs at $\lambda^0 = 1$.

The dual problem $\max\{q(\lambda) \,|\, \lambda \geq 0\}$ has the optimal solution $\lambda^0 = 1$ (see Figure 10.5). Inserting λ^0 into (10.16), we end up with the optimal point in Figure 10.2.

10.5   An interior point method

In this section we will introduce a fundamental tool in numerical algorithms for solving constrained optimization problems. The idea is to work with interior points in the constraint set and approximate an optimal solution on its boundary from the interior. This is in contrast to the Karush-Kuhn-Tucker conditions, where one essentially works on the boundary.

Interior point methods were introduced by von Neumann. They were popular in the 1960s and then lay dormant for a period of two decades until they were brought back to mainstream optimization after the Karmakar[4] algorithm for linear programming.

We will not enter into the finer numerical or mathematical details here. For this the reader is referred to the survey article [Wright (2005)] and the references there. The point is to explain and derive a simple interior point algorithm ready for immediate implementation on a computer (the author's implementation in *Mathematica* works very well on small examples).

Consider a strictly feasible optimization problem

$$\min\{f(x) \,|\, x \in S\} \tag{10.18}$$

with $S := \{x \in \mathbb{R}^n \,|\, g_1(x) \leq 0, \ldots, g_m(x) \leq 0\}$, where it is assumed that all

---

[4]N. Karmarkar (1957–). Indian mathematician.

involved functions are differentiable and convex. We will also assume that the constraint set S is compact. This ensures that (10.18) has an optimal solution.

In short, one uses the so-called logarithmic *barrier function* given as

$$B(x) = -\sum_{i=1}^{m} \log(-g_i(x))$$

to emulate the constraints. This is a convex function defined on the convex subset

$$S^o := \{x \mid g_1(x) < 0, \ldots, g_m(x) < 0\} \subseteq S,$$

which is non-empty because of our assumption of strict feasibility. The idea is to consider the family

$$f_\varepsilon(x) = f(x) + \varepsilon B(x)$$

of convex functions for $\varepsilon > 0$. A global minimum of f_ϵ is attained in a point $x_\varepsilon \in S^o$, since $B(x)$ approaches infinity when a $g_i(x)$ is close to 0 for $x \in S^o$. Under these assumptions the following result shows that the optimal value of (10.18) can be approximated by $f(x_\epsilon)$ as $\epsilon \to 0$.

Theorem 10.11. *Let x_ϵ be a point in S^o with*

$$f_\epsilon(x_\epsilon) = \min\{f_\epsilon(x) \mid x \in S^o\}$$

for $\epsilon > 0$ and $f^* = \min\{f(x) \mid x \in S\}$. Then

$$0 \le f(x_\varepsilon) - f^* \le \varepsilon m$$

and $f(x_\varepsilon) \to f^*$ as $\varepsilon \to 0$. If (10.18) has a unique optimum x^*, then by using $\epsilon = \frac{1}{n}$ we obtain a sequence $x_{\frac{1}{n}} \to x^*$ as $n \to \infty$.

Proof. Since $x_\varepsilon \in S^o$ is an interior minimum for $\varepsilon > 0$, we must have

$$0 = \nabla f_\varepsilon(x_\varepsilon) = \nabla f(x_\varepsilon) - \sum \frac{\varepsilon}{g_i(x_\varepsilon)} \nabla g_i(x_\varepsilon).$$

So x_ε is also a minimum for the (Lagrangian) function

$$f(x) + \lambda_1 g_1(x) + \cdots + \lambda_m g_m(x),$$

where

$$\lambda_i = -\frac{\varepsilon}{g_i(x_\varepsilon)} > 0.$$

Since $g_i(x) \leq 0$ for $x \in S$, we must have

$$
\begin{aligned}
f^* &\geq \min\{f(x) + \lambda_1 g_1(x) + \cdots + \lambda_m g_m(x) \mid x \in S\} \\
&= f(x_\varepsilon) + \lambda_1 g_1(x_\varepsilon) + \cdots + \lambda_m g_m(x_\varepsilon) \\
&= f(x_\varepsilon) - \varepsilon m
\end{aligned}
$$

showing that $f(x_\varepsilon) \to f^*$. If (10.18) has a unique optimum x^*, there exists for a given $\epsilon > 0$ a $\delta > 0$ by the compactness of S, such that $|f(x) - f^*| \geq \delta$ for every $x \in S$ with $|x - x^*| \geq \epsilon$. This shows that $x_{\frac{1}{n}} \to x^*$ for $n \to \infty$. □

10.5.1 Newtonian descent, exact line search and bisection

To apply the result in Theorem 10.11 we need a method for computing a minimum $x_\epsilon \in S^o$ for f_ϵ. By Corollary 9.2, this can be done solving

$$\nabla f_\epsilon(x) = 0. \tag{10.19}$$

We will use a variant of the Newton-Raphson method (see §8.1.1) for this. For a given point x_0, one step of the Newton-Raphson method for (10.19) gives the new point

$$x_0 + v, \tag{10.20}$$

where

$$v = -\nabla^2 f_\epsilon(x_0)^{-1} \nabla f_\epsilon(x_0).$$

Here it is imperative that the Hessian $\nabla^2 f_\epsilon(x_0)$ is non-singular for $\epsilon > 0$. This will always be true in the cases we consider. The *modified Newton-Raphson method* (see §5.4 in [Stoer and Bulirsch (1980)]) consists in selecting a step length $t > 0$ and replacing (10.20) by

$$x_0 + tv.$$

Usually it is a complicated issue to determine how far one should go in the direction of v. A natural choice is to use $t_0 > 0$, where t_0 minimizes the function $f(x_0 + tv)$. This technique is called *exact line search*.

Consider the one variable function

$$g(t) = f'_\epsilon(x_0 + tv) = \nabla f_\epsilon(x_0 + tv)v.$$

This is an increasing function, since $t \mapsto f_\epsilon(x_0 + tv)$ is a convex function (see Theorem 7.19). We will find a minimum of $f(x_0 + tv)$ by solving $g(t) =$

0 using a modification of the classical *bisection method*. To do this, it is necessary to first compute how far one can go in the direction of v

$$t_0 := \sup\{t \geq 0 \,|\, x_0 + tv \in S^o\} \qquad (10.21)$$

before hitting the boundary of S. The algorithm for solving $g(t) = 0$ now turns into an initial step for finding $0 < t_1 < t_2$ with $g(t_1) < 0$ and $g(t_2) > 0$ followed by classical bisection.

The initial step involves putting $\delta := t_0/2$ and $t_1 := t_2 := 0$ and then

repeating

$$t_1 := t_2; \quad t_2 := t_2 + \delta;$$
$$\delta := \delta/2;$$

as long as $g(t_2) < 0$.

After the initial step, the classical bisection algorithm is applied to $t_1 < t_2$ with $g(t_1) < 0$ and $g(t_2) > 0$. This amounts to

repeating

$$t := t_1 + \delta;$$
if $g(t) < 0$ **then**
$$t_1 := t;$$
else
$$t_2 := t;$$
end if
$$\delta := \delta/2;$$

until δ is below a small preset positive limit.

The algorithm stops with an approximate minimum for f_ϵ, when $|\nabla f_\epsilon(x_1)|$ is below a small preset positive limit.

To find f^* one typically starts with a fixed $\epsilon > 0$ and computes successive minima

$$x_{\epsilon_0}, x_{\epsilon_1}, x_{\epsilon_2}, \ldots \qquad \text{for } f_{\epsilon_0}, f_{\epsilon_1}, f_{\epsilon_2}, \ldots \qquad (10.22)$$

with $\epsilon_0 = \epsilon$ and $\epsilon_{j+1} = k\epsilon_j$ for some $0 < k < 1$ using x_{ϵ_i} as input for the modified Newton method in solving

$$\nabla f_{\epsilon_{i+1}}(x) = 0.$$

By Theorem 10.11, $f(x_{\epsilon_n}) \to f^*$ for $n \to \infty$.

10.5.2 *Polyhedral constraints*

In the case of *polyhedral constraints*, $S = \{x \mid Ax \leq b\}$, (10.21) can be computed as

$$t_0 = \min\left\{\frac{b_i - a_i x_0}{a_i v} \,\middle|\, a_i v > 0\right\},$$

where $a_1, \ldots, a_m$ are the row vectors of A and $b = (b_1, \ldots, b_m)$. Furthermore,

$$\nabla f_\epsilon(x) = \nabla f(x) + \epsilon \left(\frac{1}{b - Ax}\right)^t A$$

$$\nabla^2 f_\epsilon(x) = \nabla^2 f(x) + \epsilon A^t \operatorname{diag}\left(\frac{1}{(b - Ax)^2}\right) A$$

for $x \in S^o = \{x \mid Ax < b\}$, where the following notation is used: $1/v$ and v^2 denote the vectors with coordinates $1/v_i$ and v_i^2 respectively and

$$\operatorname{diag}(v) := \begin{pmatrix} v_1 & 0 & \cdots & 0 \\ 0 & v_2 & \cdots & 0 \\ \vdots & \vdots & \ddots & \vdots \\ 0 & 0 & \cdots & v_m \end{pmatrix}$$

for $v = (v_1, \ldots, v_m) \in \mathbb{R}^m$. The input for the minimization of f_ϵ with $\epsilon > 0$ is

  – A, b and $x_0 \in \mathbb{R}^n$ with $Ax_0 < b$.
  – The gradient function $x \mapsto \nabla f(x)$ and the Hessian $x \mapsto \nabla^2 f(x)$.

We give a few basic examples of convex optimization problems with polyhedral constraints solved numerically with the above algorithm. We begin with the linear program depicted in Figure 5.5.

Example 10.12. The polygon P sketched in Figure 5.5 has vertices $(0,0), (2,-1), (3,2), (1,2), (\frac{1}{4}, \frac{3}{2})$ and $(\frac{7}{2}, 0)$. The linear program there was $\max\{2x + \frac{5}{2}y \mid (x,y) \in P\}$, which converts into the minimization problem

$$\min\{-2x - \tfrac{5}{2}y \mid (x,y) \in P\}. \tag{10.23}$$

Figure 10.6 illustrates convergence to the optimal point $(3,2)$ in (10.23) for $\epsilon = 16, 8, 4, 2, 1, 0.5, 0.2$ in (10.22).

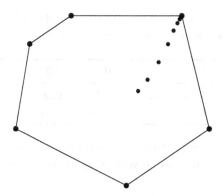

Figure 10.6: Convergence of the interior point method for a linear program.

Example 10.13. In Figure 10.1, the smallest circle containing seven given points was sketched. These points were

$$(0,0), (2,2), (-3,2), (1,0), (-2,1), (-1,3) \quad \text{and} \quad (0,4).$$

The center (x, y) and radius r of the smallest circle can be found minimizing the convex function $f(x, y, \lambda) = x^2 + y^2 + 2\lambda$, subject to the polyhedral constraints

$$\tfrac{1}{2}(x_i^2 + y_i^2) \le x_i x + y_i y + \lambda \qquad \text{for } i = 1, \ldots, n,$$

with $r^2 = x^2 + y^2 + 2\lambda$. We need the constraint set to be bounded to guarantee that the interior point algorithm will work. This is done by adding a suitable upper bound on λ (here $\lambda \le 10$),

$$
\begin{aligned}
\lambda &\le 10 \\
-\lambda &\le 0 \\
-2x - 2y - \lambda &\le -4 \\
3x - 2y - \lambda &\le -6.5 \\
-x \qquad - \lambda &\le -0.5 \\
2x - y - \lambda &\le -2.5 \\
x - 3y - \lambda &\le -5 \\
-4y - \lambda &\le -8.
\end{aligned}
$$

Starting with $x_0 = (0, 0, 9)$ and $\epsilon = 1$ here is a table with computations of minima for f_ϵ for seven decreasing values of $\epsilon > 0$.

ϵ	1.00	0.5	0.2	0.1	0.05	0.01	0.0001
x	−0.44	−0.45	−0.49	−0.50	−0.50	−0.50	−0.50
y	1.85	1.81	1.82	1.84	1.87	1.93	1.99

The table strongly suggests that the minimal circle containing the points has center $(-0.5, 2)$. This can be verified by other means.

Example 10.14.

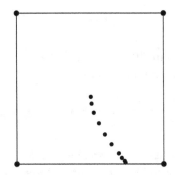

Figure 10.7: Convergence to optimal point $(0.5, -1)$ for $\epsilon = 8, 4, 2, \ldots, 0.05$ in $\min\{x^2 + y^2 + xy + 2y \mid -1 \leq x \leq 1, -1 \leq y \leq 1\}$.

Consider the optimization problem

$$\min\{f(x, y) \mid -1 \leq x \leq 1, -1 \leq y \leq 1\}$$

with $f(x, y) = x^2 + y^2 + xy + 2y$. Starting with $x_0 = (0, 0)$ and $\epsilon = 8$ we get the table

ϵ	8	4	2	1	0.5	0.25	0.13	0.05
x	0.00	0.02	0.05	0.12	0.20	0.29	0.36	0.43
y	−0.11	−0.20	−0.32	−0.46	−0.61	−0.74	−0.83	−0.92

indicating convergence to the optimal point $(0.5, -1)$ on the boundary of the square (see Figure 10.7).

10.6 Maximizing convex functions over polytopes

Maximizing a convex function over a polytope turns out to be easier than the corresponding minimization problem. This is made precise in the following result.

Theorem 10.15. *Let $V = \{x_1, \ldots, x_m\} \subseteq \mathbb{R}^n$ and*

$$P = \text{conv}(\{x_1, \ldots, x_m\})$$
$$= \{\lambda_1 x_1 + \cdots + \lambda_m x_m \mid \lambda_i \geq 0, \ \lambda_1 + \cdots + \lambda_m = 1\}.$$

The optimization problem

$$\max\{f(x) \mid x \in P\}$$

has a global maximum at $x_0 \in V$, where

$$f(x_0) = \max\{f(x_1), \ldots, f(x_m)\}.$$

This maximum is unique if and only if $f(x_0) = f(x_j)$ for a unique $j = 1, \ldots, m$.

Proof. For $x = \lambda_1 x_1 + \cdots + \lambda_m x_m \in P$ we have

$$f(x) \leq \lambda_1 f(x_1) + \cdots + \lambda_m f(x_m), \tag{10.24}$$

by Corollary 7.4. The right hand side of (10.24) satisfies

$$\lambda_1 f(x_1) + \cdots + \lambda_m f(x_m) \leq \lambda_1 M + \cdots + \lambda_m M = M,$$

where $M = \max\{f(x_1), \ldots, f(x_m)\}$, since $\lambda_1, \ldots, \lambda_m \geq 0$ and $\lambda_1 + \cdots + \lambda_m = 1$. We may now put $x_0 = x_j$, where $f(x_j) = M$. If the maximum is unique, there can only be one $j = 1, \ldots, m$ with $f(x_j) = f(x_0)$. If there is only one such j, then

$$f(\lambda_1 x_1 + \cdots + \lambda_j x_j + \cdots + \lambda_m x_m)$$
$$\leq \lambda_1 f(x_1) + \cdots + \lambda_j f(x_j) + \cdots + \lambda_m f(x_m)$$
$$< \lambda_1 f(x_j) + \cdots + \lambda_j f(x_j) + \cdots + \lambda_m f(x_j) = f(x_j)$$

if and only if $\lambda_j < 1$, where $\lambda_1 + \cdots + \lambda_m = 1$ and $\lambda_i \geq 0$. This proves the uniqueness of the maximum x_j. $\square$

Remark 10.16. Notice the strong difference to the minimization problem

$$\min\{f(x) \mid x \in P\}.$$

Here you can take any polygon in the plane containing $Q = (0,0)$ in its interior and the convex function $f(x,y) = x^2 + y^2$. Clearly f is uniquely minimal at $Q \in P$ and Q is not a vertex of P. However, if f is a linear function i.e., $f(x) = \alpha^t x$ for some $\alpha \in \mathbb{R}^n$, then $-f(x)$ is also convex and

$$\min\{f(x) \mid x \in P\} = -\max\{-f(x) \mid x \in P\}.$$

10.6.1   *Convex functions are continuous on open subsets*

We will give a purely mathematical application of the idea in the proof of Theorem 10.15. If you go back and look at the proof of Theorem 7.7 you will realize that the proof given below is a natural generalization to higher dimensions (using boxes instead of closed intervals). I am grateful to Jørgen Vesterstrøm for explaining this proof to me.

Theorem 10.17. *Let $f : U \to \mathbb{R}$ be a convex function, where $U \subseteq \mathbb{R}^n$ is an open convex subset. Then f is a continuous function on U.*

Proof. We will prove that f is continuous at $x_0 \in U$. Here we may assume $x_0 = 0$ and $f(x_0) = 0$ by considering the convex function $g(x) = f(x + x_0) - f(x_0)$ defined on the convex open subset $\{x - x_0 \mid x \in U\}$ containing 0. For sufficiently small $\delta > 0$,

$$S = \delta \operatorname{conv}(\{\pm e_1, \dots, \pm e_n\})$$
$$= \operatorname{conv}(\{\pm \delta e_1, \dots, \pm \delta e_n\}) \subseteq U,$$

where $e_1, \dots, e_n$ are the canonical basis vectors of $\mathbb{R}^n$. Geometrically we are scaling the box $\operatorname{conv}(\{\pm e_1, \dots, \pm e_n\})$ so that it fits inside a small ball around $0 \in U$. Since

$$S = \{\pm \mu_1 e_1 \pm \dots \pm \mu_n e_n \mid \mu_i \geq 0, \mu_1 + \dots + \mu_n \leq \delta\}$$

it follows for $v = (x_1, \dots, x_n) \in S$ that

$$f(v) = f(x_1 e_1 + \dots + x_n e_n)$$
$$= f\left(\frac{|x_1|}{\delta} \operatorname{sgn}(x_1) \delta e_1 + \dots + \frac{|x_n|}{\delta} \operatorname{sgn}(x_n) \delta e_n + \mu \cdot 0\right)$$
$$\leq (|x_1| + \dots + |x_n|)\frac{M}{\delta}$$

by Jensen's inequality (see §7.2), where

$$M = \max\{f(\pm \delta e_i) \mid i = 1, \ldots, n\}$$

and $0 \in S$ is picking up the lag

$$\mu = 1 - \frac{|x_1|}{\delta} - \cdots - \frac{|x_n|}{\delta} \geq 0$$

in the convex combination. Using that $-f(-x) \leq f(x)$ for $x \in S$ and $-S = S$ we get the lower bound needed in deducing that

$$|f(v)| \leq (|x_1| + \cdots + |x_n|)\frac{M}{\delta} \tag{10.25}$$

for every $v = (x_1, \ldots, x_n) \in S$. Since S contains the ball $B(0, \epsilon)$ for $\epsilon = \frac{\delta}{n} > 0$ it follows by (10.25) that the sequence $(f(v_m))$ must converge to $0 = f(0)$ if the sequence $(v_m) \subseteq U$ converges to 0: f is continuous at 0. □

10.7 Exercises

Exercise 10.1. Does a convex function $f : \mathbb{R} \to \mathbb{R}$ with a unique global minimum have to be strictly convex? What if f is differentiable?

Exercise 10.2. Let $f : \mathbb{R}^2 \to \mathbb{R}$ be a differentiable convex function and

$$S = \{(x, y) \mid -1 \leq x \leq 2, -1 \leq y \leq 1\}.$$

Suppose that $\nabla f(x_0) = (1, 0)$ for $x_0 = (-1, \frac{1}{2})$. Prove that x_0 is a minimum for f defined on S.

Exercise 10.3. Guess the solution to the optimization problem

$$\min\{(x - 5)^2 + (y - 5)^2 \mid x \geq 0, y \geq 0, x^2 + y^2 \leq 25\}.$$

Show that your guess was correct!

Exercise 10.4. Let

$$S = \left\{(x, y) \in \mathbb{R}^2 \, \middle| \, \begin{matrix} -x - y \leq 0 \\ 2x - y \leq 1 \\ -x + 2y \leq 1 \end{matrix} \right\}.$$

(i) Use the KKT conditions to solve the minimization problem

$$\min\{-x - 4y \,|\, (x, y) \in S\}.$$

(ii) Use the KKT conditions to solve the minimization problem

$$\min\{x + y \,|\, (x, y) \in S\}.$$

(iii) Solve the optimization problem

$$\max\{(x + 1)^2 + (y + 1)^2 \,|\, (x, y) \in S\}.$$

Give a geometric interpretation of your answer.

Exercise 10.5. Solve the optimization problem

$$\min\left\{x^2 + 2y^2 + 3z^2 - 2xz - xy \,\middle|\, \begin{matrix} 2x^2 + y^2 + z^2 \le 4 \\ 1 \ge x + y + z \end{matrix}\right\}.$$

Exercise 10.6. Let $S = \{(x, y) \,|\, 2x^2 + y^2 \le 3,\ x^2 + 2y^2 \le 3\}$ and $f(x, y) = (x - 4)^2 + (y - 4)^2$.

(i) State the KKT conditions for $\min\{f(x, y) \,|\, (x, y) \in S\}$ for $(x, y) = (1, 1)$.

(ii) Suppose now that $g(x, y) = (x - a)^2 + (y - b)^2$. For which a and b does $\min\{g(x, y) \,|\, (x, y) \in S\}$ have optimum in $(1, 1)$? State the KKT conditions when $(a, b) = (1, 1)$.

Exercise 10.7. Let $f : \mathbb{R}^2 \to \mathbb{R}$ be given by

$$f(x, y) = (x - 1)^2 + (y - 1)^2 + 2xy.$$

(i) Show that f is a convex function.

(ii) Find $\min\{f(x, y) \,|\, (x, y) \in \mathbb{R}^2\}$. Is this minimum unique? Is f a strictly convex function.

Let

$$S = \{(x, y) \in \mathbb{R}^2 \,|\, x + y \le 0,\quad x - y \le 0\}.$$

(iii) Apply the KKT-conditions to decide if $(-1, -1)$ is an optimal solution to

$$\min\{f(x, y) \,|\, (x, y) \in S\}.$$

(iv) Find

$$m = \min\{f(x, y) \,|\, (x, y) \in S\}$$

and

$$\{(x, y) \in \mathbb{R}^2 \,|\, f(x, y) = m\}.$$

Exercise 10.8. Let $f : \mathbb{R}^2 \to \mathbb{R}$ be given by

$$f(x,y) = x^2 + y^2 - e^{x-y-1}$$

and let

$$C = \{(x,y) \,|\, x - y \leq 0\}.$$

(i) Show that $f : \mathbb{R}^2 \to \mathbb{R}$ is not a convex function.

(ii) Show that f is a convex function on the open subset

$$\{(x,y) \in \mathbb{R}^2 \,|\, x - y < \tfrac{1}{2}\}$$

and conclude that f is convex on C.

(iii) Show that $v = (0,0)$ is an optimal solution for the optimization problem $\min\{f(v) \,|\, v \in C\}$. Is v a unique optimal solution here?

Exercise 10.9. Let $f : \mathbb{R}^4 \to \mathbb{R}$ be given by

$$f(x_1, x_2, x_3, x_4) = (x_1 - x_3)^2 + (x_2 - x_4)^2$$

and $C \subseteq \mathbb{R}^4$ by

$$C = \{(x_1, x_2, x_3, x_4) \in \mathbb{R}^4 \,|\, x_1^2 + (x_2 - 2)^2 \leq 1, \, x_3 - x_4 \geq 0\}.$$

(i) Show that f is a convex function. Is f strictly convex?

(ii) Show that C is a convex subset of $\mathbb{R}^4$.

(iii) Does there exist an optimal point $v = (x_1, x_2, x_3, x_4) \in \mathbb{R}^4$ for the minimization problem

$$\min_{v \in C} f(v)$$

with $x_3 = x_4 = 0$?

(iv) Does there exist an optimal point $v = (x_1, x_2, x_3, x_4) \in \mathbb{R}^4$ for the minimization problem

$$\min_{v \in C} f(v)$$

with $x_3 = x_4 = 1$?

Exercise 10.10. Let

$$f(x,y) = (x - 1)^2 + y^2$$

and

$$C = \{(x,y) \in \mathbb{R}^2 \,|\, -1 \leq x \leq 0, \, -1 \leq y \leq 1\}.$$

(i) Solve the optimization problem

$$\min\{f(v) \,|\, v \in C\}.$$

(ii) Solve the optimization problem

$$\max\{f(v) \,|\, v \in C\}.$$

How many optimal solutions are there? Why?

Exercise 10.11. Let $f : \mathbb{R}^2 \to \mathbb{R}$ be given by

$$f(x, y) = \tfrac{1}{2}x^2 + y^2 - 2y + 2.$$

Below, the minimization problem

$$\min\{f(x, y) \,|\, (x, y) \in S\} \qquad\qquad (10.26)$$

is analyzed for various subsets $S \subseteq \mathbb{R}^2$.

(i) Show that f is a convex function
(ii) Show that $h(x, y) = -x + 2y - 1$ is a convex and concave function.
(iii) Let

$$S = \{(x, y) \in \mathbb{R}^2 \,|\, -x + 2y \le 1\}.$$

Show that $(-1, 0) \in S$ cannot be an optimal solution to (10.26). Find an optimal solution to (10.26).
(iv) Find an optimal solution in (10.26) for

$$S = \{(x, y) \in \mathbb{R}^2 \,|\, -x + 2y \ge 1\}.$$

(v) Are the optimal solutions in (iii) and (iv) unique?

Exercise 10.12. Let T denote the convex hull of $(1, 1), (-1, 2), (2, 3) \in \mathbb{R}^2$.

(i) Solve the optimization problem

$$\max\{x^2 + y^3 \,|\, (x, y) \in T\}.$$

Is your solution unique?
(ii) Solve the optimization problem

$$\min\{x^2 + y^2 \,|\, (x, y) \in T\}$$

and give a geometric interpretation.

Exercise 10.13. Let C denote the set of points $(x, y) \in \mathbb{R}^2$ such that

$$x^2 + 2y^2 \leq 1$$
$$x + y \geq 1$$
$$y \leq x.$$

(i) Show that C is a convex set and that there exists $(x_0, y_0) \in \mathbb{R}^2$ such that

$$x_0^2 + 2y_0^2 < 1$$
$$x_0 + y_0 > 1$$
$$y_0 < x_0.$$

(ii) Solve the optimization problem

$$\max\{x + 3y \mid (x, y) \in C\}.$$

Is your solution unique?

Exercise 10.14. Let $f : \mathbb{R}^2 \to \mathbb{R}$ be given by

$$f(x, y) = x^2 + y^4 + xy.$$

(i) Is f a convex function?
(ii) Let F denote the convex hull of $(-1, 1)$, $(-1, 2)$, $(-2, 2)$, $(-3, 1) \in \mathbb{R}^2$. Solve the optimization problem

$$\max\{f(x, y) \mid (x, y) \in F\}.$$

Is your maximum unique?
(iii) Show in detail that

$$\min\{f(x, y) \mid (x, y) \in F\}$$

has the unique solution $x = -1$ og $y = 1$.
(iv) Does f have a unique global minimum on $\mathbb{R}^2$?

Exercise 10.15. Let $f : \mathbb{R}^2 \to \mathbb{R}$ be given by

$$f(x, y) = x^2 + xy + y^2$$

and let

$$S = \left\{ (x, y) \in \mathbb{R}^2 \, \middle| \, \begin{array}{l} y - x \geq 1 \\ y + x \geq 1 \\ y \quad\;\; \leq 2 \end{array} \right\}.$$

(i) Is $(0,1)$ an optimal solution to

$$\min\{f(x,y)\,|\,(x,y)\in S\}?$$

(ii) Find

$$\max\{f(x,y)\,|\,(x,y)\in S\}.$$

Exercise 10.16. Let $f:\mathbb{R}^2\to\mathbb{R}$ be given by

$$f(x,y)=2x^2+3x+y^2+y.$$

(i) Show that f is a convex function and solve the minimization problem $\min\{f(x,y)\,|\,x,y\in\mathbb{R}\}$.

Now let

$$S=\left\{(x,y)\in\mathbb{R}^2\,\middle|\,\begin{array}{c}x^2+(y+1)^2\le 1\\ y-x\le 0\end{array}\right\}$$

and consider the minimization problem (P) given by

$$\min\{f(x,y)\,|\,(x,y)\in S\}.$$

(ii) Show using the KKT conditions that $(0,0)$ is not optimal for (P).

(iii) Find an optimal solution for (P). Is it unique?

Exercise 10.17. Let $f:\mathbb{R}^2\to\mathbb{R}$ be a differentiable function and suppose that (x_0,y_0) is a saddle point of f in the sense that

$$f(x_0,y_0)\le f(x,y_0)\qquad\text{for every }x\in\mathbb{R}$$

$$f(x_0,y)\le f(x_0,y_0)\qquad\text{for every }y\in\mathbb{R}.$$

Prove that (x_0,y_0) is a critical point for f. Show that a local extremum for f is not a saddle point. Is a critical point for f, which is not an extremum a saddle point in the above sense?

Exercise 10.18. Pick two optimization problems from the exercises in this section and compute their dual problems. Solve the dual problems and compare with the solution of the original problems.

Exercise 10.19. Go through the steps of the interior point algorithm in §10.5 for the basic example $\min\{x\,|\,0\le x\le 1\}$.

Exercise 10.20. Consider the setup in §10.5. Prove that f_ϵ is a convex function on S^o for $\epsilon>0$. Is the Hessian of f_ϵ always positive definite?

Exercise 10.21. Implement the interior point algorithm in §10.5 in your favorite language (C, Haskell, *Mathematica*, ...). Test your implementation on the examples in §10.5.

Appendix A

Analysis

In this appendix we give a very brief overview of the basic concepts of introductory mathematical analysis. Focus is directed at building things from scratch with applications to convex sets. We have not formally constructed the real numbers.

A.1 Measuring distances

The limit concept is a cornerstone in mathematical analysis. We need a formal way of stating that two vectors are far apart or close together.

The length of a vector $x = (x_1, \ldots, x_n) \in \mathbb{R}^n$ is defined as

$$|x| = \sqrt{x_1^2 + \cdots + x_n^2}.$$

Our first result about the length is the following lemma called the inequality of Cauchy-Schwarz. It was discovered by Cauchy in 1821 and rediscovered by Schwarz[1] in 1888.

Lemma A.1. *For $x = (x_1, \ldots, x_n) \in \mathbb{R}^n$ and $y = (y_1, \ldots, y_n) \in \mathbb{R}^n$ the inequality*

$$(x^t y)^2 = (x_1 y_1 + \cdots + x_n y_n)^2 \leq (x_1^2 + \cdots + x_n^2)(y_1^2 + \cdots + y_n^2) = |x|^2 |y|^2$$

holds. If

$$(x^t y)^2 = (x_1 y_1 + \cdots + x_n y_n)^2 = (x_1^2 + \cdots + x_n^2)(y_1^2 + \cdots + y_n^2) = |x|^2 |y|^2,$$

then x and y are proportional i.e., $x = \lambda y$ for some $\lambda \in \mathbb{R}$.

---

[1] Hermann Amandus Schwarz (1843–1921), German mathematician.

Proof.   For $n = 2$ you can explicitly verify that

$$(x_1^2 + x_2^2)(y_1^2 + y_2^2) - (x_1y_1 + x_2y_2)^2 = (x_1y_2 - y_1x_2)^2. \qquad \text{(A.1)}$$

This proves the inequality for $n = 2$. If equality holds, we must have

$$x_1y_2 - y_1x_2 = \begin{vmatrix} x_1 & y_1 \\ x_2 & y_2 \end{vmatrix} = 0.$$

This implies as you can check that there exists $\lambda \in \mathbb{R}$ such that $x_1 = \lambda y_1$ and $x_2 = \lambda y_2$.

The formula in (A.1) generalizes for $n > 2$ by induction (Exercise A.1) to

$$(x_1^2 + \cdots + x_n^2)(y_1^2 + \cdots + y_n^2) - (x_1y_1 + \cdots + x_ny_n)^2 \qquad \text{(A.2)}$$
$$= (x_1y_2 - y_1x_2)^2 + \cdots + (x_{n-1}y_n - y_{n-1}x_n)^2,$$

where the last sum is over the squares of the 2×2 minors in the matrix

$$A = \begin{pmatrix} x_1 & x_2 & \cdots & x_{n-1} & x_n \\ y_1 & y_2 & \cdots & y_{n-1} & y_n \end{pmatrix}.$$

The formula in (A.2) proves the inequality. If

$$(x_1^2 + \cdots + x_n^2)(y_1^2 + \cdots + y_n^2) = (x_1y_1 + \cdots + x_ny_n)^2,$$

then (A.2) shows that all the 2×2-minors in A vanish. The existence of λ giving proportionality between x and y is deduced as for $n = 2$.   □

If you know about the vector (cross) product $u \times v$ of two vectors $u, v \in \mathbb{R}^3$ you will see that the idea in the above proof comes from the formula

$$|u|^2|v|^2 = |u^tv|^2 + |u \times v|^2.$$

A different and perhaps more well known proof of Lemma A.1 is outlined in Exercise A.2.

Theorem A.2. *For two vectors $x, y \in \mathbb{R}^n$ the inequality*

$$|x + y| \le |x| + |y|$$

holds.

Proof. Lemma A.1 shows that

$$|x + y|^2 = (x + y)^t(x + y)$$
$$= |x|^2 + |y|^2 + 2x^t y$$
$$\leq |x|^2 + |y|^2 + 2|x||y|$$
$$= (|x| + |y|)^2$$

proving the inequality. $\qquad\square$

The distance between two points in $\mathbb{R}^n$ is defined to be the length of the line segment between them.

Definition A.3. The distance between x and y in $\mathbb{R}^n$ is defined as

$$|x - y|.$$

From Theorem A.2 you get formally

$$|x - z| = |x - y + y - z| \leq |x - y| + |y - z|$$

for $x, y, z \in \mathbb{R}^n$. This is the triangle inequality for distance saying that the shorter way is always along the diagonal instead of the other two sides in a triangle, see Figure A.1.

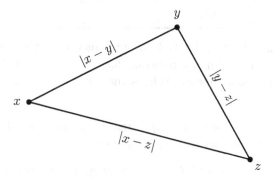

Figure A.1: The triangle inequality.

A.2 Sequences

Limits appear in connection with (infinite) sequences of vectors in $\mathbb{R}^n$. We need to formalize this.

Definition A.4. A sequence in $\mathbb{R}^n$ is a function $f : \{1, 2, \dots\} \to \mathbb{R}^n$. A subsequence f_I of f is f restricted to an infinite subset $I \subseteq \{1, 2, \dots\}$.

A sequence f is usually denoted by an infinite tuple $(x_n) = (x_1, x_2, \dots)$, where $x_n = f(n)$. A subsequence of (x_n) is denoted (x_{n_i}), where $I = \{n_1, n_2, \dots\}$ and $n_1 < n_2 < \cdots$. A subsequence f_I is in itself a sequence, since it is given by picking out an infinite subset I of $\{1, 2, \dots\}$ and then letting $f_I(j) = f(n_j) = x_{n_j}$.

This definition is quite formal. Once you get to work with it, you will discover that it is easy to handle. In practice, sequences are often listed as

$$1, \ 2, \ 3, \ 4, \ 5, \ 6, \ \dots \tag{A.3}$$

$$2, \ 4, \ 6, \ 8, \ 10, \ \dots \tag{A.4}$$

$$2, \ 6, \ 4, \ 8, \ 10, \ \dots \tag{A.5}$$

$$1, \ \tfrac{1}{2}, \ \tfrac{1}{3}, \ \tfrac{1}{4}, \ \dots \tag{A.6}$$

Formally these sequences are given in the table below

	x_1	x_2	x_3	x_4	x_5	x_6	$\cdots$
(A.3)	1	2	3	4	5	6	$\cdots$
(A.4)	2	4	6	8	10	12	$\cdots$
(A.5)	2	6	4	8	10	12	$\cdots$
(A.6)	1	$\tfrac{1}{2}$	$\tfrac{1}{3}$	$\tfrac{1}{4}$	$\tfrac{1}{5}$	$\tfrac{1}{6}$	$\cdots$

The sequence (z_n) in (A.4) is a subsequence of the sequence (x_n) in (A.3). You can see this by noticing that $z_n = x_{2n}$ and checking with the definition of a subsequence. Why is the sequence in (A.5) not a subsequence of (x_n)?

Definition A.5. A sequence (x_n) of real numbers is called *increasing* if $x_1 \leq x_2 \leq \cdots$ and *decreasing* if $x_1 \geq x_2 \geq \cdots$.

The sequences (A.3) and (A.4) are increasing. The sequence (A.6) is decreasing, whereas (A.5) is neither increasing nor decreasing.

You probably agree that the following lemma is very intuitive.

Lemma A.6. *Let T be an infinite subset of $\{1, 2, \dots\}$ and F a finite subset. Then $T \setminus F$ is infinite.*

However, infinity should be treated with the greatest respect in this setting. It sometimes leads to really surprising statements such as the following.

Lemma A.7. *A sequence (x_n) of real numbers always contains an increasing or a decreasing subsequence.*

Proof. We will prove that if (x_n) does not contain an increasing subsequence, then it must contain a decreasing subsequence (x_{n_i}), with

$$x_{n_1} > x_{n_2} > x_{n_3} > \cdots$$

The key observation is that if (x_n) does not contain an increasing subsequence, then there exists N_0 such that $X_{N_0} > x_n$ for every $n > N_0$. If this was not so, (x_n) would contain an increasing subsequence. You can try this out yourself!

The first element in our subsequence will be X_{N_0}. In the same way there exists $N_1 > N_0$ such that $x_n < X_{N_1}$ for $n > N_1$. We let the second element in our subsequence be x_{N_1} and so on. We use nothing but Lemma A.6 in this process. If the process should come to a halt after a finite number of steps $x_{n_1}, x_{n_2}, \ldots, x_{n_k}$, then the sequence (x_j) with $j \geq n_k$ must contain an increasing subsequence, which is also an increasing subsequence of (x_n). This is a contradiction. □

Definition A.8. A sequence (x_n) *converges to x* (this is written $x_n \to x$) if

$$\forall \epsilon > 0 \, \exists N \in \mathbb{N} \, \forall n \geq N : |x - x_n| \leq \epsilon.$$

Such a sequence is called *convergent*.

This is a very formal (but necessary!) way of expressing that ...

> *the bigger n gets, the closer x_n is to x.*

You can see that (A.3) and (A.4) are not convergent, whereas (A.6) converges to 0. To practice the formal definition of convergence you should (Exercise A.4) prove the following proposition.

Proposition A.9. *Let (x_n) and (y_n) be sequences in $\mathbb{R}^n$. Then the following hold.*

(1) *If $x_n \to x$ and $x_n \to x'$, then $x = x'$.*
(2) *If $x_n \to x$ and $y_n \to y$, then*

$$x_n + y_n \to x + y \quad and \quad x_n y_n \to xy.$$

Before moving on with the more interesting aspects of convergent sequences we need a very important property of the real numbers.

A.2.1   *Supremum and infimum*

A subset $S \subseteq \mathbb{R}$ is *bounded from above* if there exists $U \in \mathbb{R}$ such that $x \leq U$ for every $x \in S$. Similarly S is *bounded from below* if there exists $L \in \mathbb{R}$ such that $L \leq x$ for every $x \in S$.

Theorem A.10. *Let $S \subseteq \mathbb{R}$ be a subset bounded from above. Then there exists a number (supremum) $\sup(S) \in \mathbb{R}$, such that*

(1) $x \leq \sup(S)$ *for every* $x \in S$.
(2) *For every* $\epsilon > 0$, *there exists* $x \in S$ *such that*

$$x > \sup(S) - \epsilon.$$

Similarly we have for a bounded below subset S that there exists a number (infimum) $\inf(S)$ such that

(3) $x \geq \inf(S)$ *for every* $x \in S$.
(4) *For every* $\epsilon > 0$, *there exists* $x \in S$ *such that*

$$x < \inf(S) + \epsilon.$$

Let $S = \{x_n \mid n = 1, 2, \ldots\}$, where (x_n) is a sequence. Then (x_n) is bounded from above from if S is bounded from above, and similarly bounded from below if S is bounded from below.

Lemma A.11. *Let (x_n) be a sequence of real numbers. Then (x_n) is convergent if*

(1) (x_n) *is increasing and bounded from above.*
(2) (x_n) *is decreasing and bounded from below.*

Proof.   In the increasing case $\sup\{x_n \mid n = 1, 2, \ldots\}$ is the limit. In the decreasing case $\inf\{x_n \mid n = 1, 2, \ldots\}$ is the limit.   □

A.3   Bounded sequences

A sequence of real numbers is called bounded if it is both bounded from above and below.

Corollary A.12. *A bounded sequence of real numbers has a convergent subsequence.*

Proof. This is a consequence of Lemmas A.7 and A.11. $\qquad\square$

We want to generalize this result to $\mathbb{R}^m$ for $m > 1$. Surprisingly this is not so hard once we use properties of infinite sets. First we need to define bounded subsets here.

A subset $S \subseteq \mathbb{R}^m$ is called *bounded* if there exists $R > 0$ such that $|x| \le R$ for every $x \in S$. This is a very natural definition. You want your set S to be contained in vectors of length bounded by R.

Theorem A.13. *A bounded sequence (x_n) in $\mathbb{R}^m$ has a convergent subsequence.*

Proof. Let the sequence be given by

$$x_n = (x_{1n}, \ldots, x_{mn}) \in \mathbb{R}^m.$$

The m sequences of coordinates $(x_{1n}), \ldots, (x_{mn})$ are all bounded sequences of real numbers. So the first one (x_{1n}) has a convergent subsequence (x_{1n_i}). Nothing is lost in replacing (x_n) with its subsequence (x_{n_i}). Once we do this we know that the first coordinate converges! Move on to the sequence given by the second coordinate and repeat the procedure. Eventually we end with a convergent subsequence of the original sequence. $\qquad\square$

A.4 Closed subsets and open subsets

Definition A.14. A subset $F \subseteq \mathbb{R}^n$ is called closed if every convergent sequence $(x_n) \subseteq F$ converging to $x \in \mathbb{R}^n$ has $x \in F$. A subset $U \subseteq \mathbb{R}^n$ is called open if its complement $\mathbb{R}^n \setminus U$ is closed.

Clearly $\mathbb{R}^n$ is closed. Also an arbitrary intersection of closed sets is closed. By definition of convergence of a sequence, a subset $U \subseteq \mathbb{R}^n$ is open if and only if for every $x_0 \in U$ there exists $\epsilon > 0$ such that

$$B(x, \epsilon) := \{x \in \mathbb{R}^n \mid |x - x_0| \le \epsilon\} \subseteq U.$$

Definition A.15. The closure of a subset $S \subseteq \mathbb{R}^n$ is defined as

$$\overline{S} = \{x \in \mathbb{R}^n \mid x_n \to x, \text{ where } (x_n) \subseteq S \text{ is a convergent sequence}\}.$$

Proposition A.16. *Let $S \subseteq \mathbb{R}^n$. Then $\overline{S}$ is closed.*

Proof.   Consider a convergent sequence $(y_m) \subseteq \overline{S}$ with $y_m \to y$. We wish to prove that $y \in \overline{S}$. By definition there exists for each y_m a convergent sequence $(x_{m,n}) \subseteq S$ with

$$x_{m,n} \to y_m.$$

For each m we pick $x_m := x_{m,n}$ for n big enough such that $|y_m - x_m| < 1/m$. We claim that $x_m \to y$. This follows from the inequality

$$|y - x_m| = |y - y_m + y_m - x_m| \le |y - y_m| + |y_m - x_m|,$$

using that $y_m \to y$ and $|y_m - x_m|$ being small for $m \gg 0$. $\square$

The following proposition comes in very handy.

Proposition A.17. *Let $F_1, \ldots, F_m \subseteq \mathbb{R}^n$ be finitely many closed subsets. Then*

$$F := F_1 \cup \cdots \cup F_m \subseteq \mathbb{R}^n$$

is a closed subset.

Proof.   Let $(x_n) \subseteq F$ denote a convergent sequence with $x_n \to x$. We must prove that $x \in F$. Again distributing infinitely many elements in finitely many boxes implies that one box must contain infinitely many elements. Here this means that at least one of the sets

$$\mathbb{N}_i = \{n \in \mathbb{N} \mid x_n \in F_i\}, \qquad i = 1, \ldots, m$$

must be infinite. If $\mathbb{N}_k$ infinite then $\{x_j \mid j \in \mathbb{N}_k\}$ is a convergent (why?) subsequence of (x_n) with elements in F_k. But F_k is closed so that $x \in F_k \subseteq F$. $\square$

A.5   The interior and boundary of a set

The interior $\text{int}(S)$ of a subset $S \subseteq \mathbb{R}^n$ consists of the elements which are not limits of sequences of elements outside S. The boundary ∂S consists of the points which can be approximated both from the inside and outside. This is formalized in the following definition.

Definition A.18. Let $S \subseteq \mathbb{R}^n$. Then the interior $\text{int}(S)$ of S is

$$\text{int}(S) = \mathbb{R}^n \setminus \overline{\mathbb{R}^n \setminus S}.$$

The boundary ∂S is

$$\partial S = \overline{S} \cap \overline{\mathbb{R}^n \setminus S}.$$

These terse definitions call for some words. The interior int(S) is the set of points in S, which are not limits of sequences coming from the complement of S. Informally they are the points which cannot be approximated arbitrarily well with points outside S.

The boundary are precisely those points that are limits of sequences of points from S and sequences from the complement of S. Informally they are the points that can be approximated arbitrarily well from both inside and outside S.

A.6 Continuous functions

Definition A.19. A function

$$f : S \to \mathbb{R}^n,$$

where $S \subseteq \mathbb{R}^m$ is called continuous if $f(x_n) \to f(x)$ for every convergent sequence $(x_n) \subseteq S$ with $x_n \to x \in S$.

We would like the length function to be continuous. This is the content of the following proposition.

Proposition A.20. *The length function $f(x) = |x|$ is a continuous function from $\mathbb{R}^n$ to $\mathbb{R}$.*

Proof. You can deduce from the triangle inequality that

$$\big| |x| - |y| \big| \le |x - y|$$

for every $x, y \in \mathbb{R}^n$. This shows that

$$|f(x) - f(x_n)| \le |x - x_n|$$

proving that $f(x_n) \to f(x)$ if $x_n \to x$. Therefore $f(x) = |x|$ is a continuous function. $\square$

The following result is very useful for proving that certain subsets are closed.

Lemma A.21. *If $f : \mathbb{R}^m \to \mathbb{R}^n$ is continuous then*

$$f^{-1}(F) = \{x \in \mathbb{R}^n \mid f(x) \in F\} \subseteq \mathbb{R}^m$$

is a closed subset, where F is a closed subset of $\mathbb{R}^n$.

Proof.   If $(x_n) \subseteq f^{-1}(F)$ with $x_n \to x$, then $f(x_n) \to f(x)$ by the continuity of f. As F is closed we must have $f(x) \in F$. Therefore $x \in f^{-1}(F)$.   □

A.7   The main theorem

A closed and bounded subset of $\mathbb{R}^n$ is called *compact*. Even though the following theorem is a bit dressed up, its applications are many and quite down to earth. Do not fool yourself by the simplicity of the proof. The proof is only simple because we have the right definitions.

Theorem A.22. *Let $f : K \to \mathbb{R}^n$ be a continuous function, where $K \subseteq \mathbb{R}^m$ is compact. Then $f(K) = \{f(x) \,|\, x \in K\}$ is compact in $\mathbb{R}^n$.*

Proof.   Suppose that $f(K)$ is not bounded. Then we may find a sequence $(x_n) \subseteq K$ such that $|f(x_n)| \geq n$. However, by Theorem A.13 we know that (x_n) has a convergent subsequence (x_{n_i}) with $x_{n_i} \to x$. Since K is closed we must have $x \in K$. The continuity of f gives $f(x_{n_i}) \to f(x)$. This contradicts our assumption that $|f(x_{n_i})| \geq n_i$ – after all, $|f(x)|$ is finite.

Proving that $f(K)$ is closed is almost the same idea: suppose that $f(x_n) \to y$. Then again (x_n) must have a convergent subsequence (x_{n_i}) with $x_{n_i} \to x \in K$. Therefore $f(x_{n_i}) \to f(x)$ and $y = f(x)$, showing that $f(K)$ is closed.   □

One of the useful consequences of this result is the following.

Corollary A.23. *Let $f : K \to \mathbb{R}$ be a continuous function, where $K \subseteq \mathbb{R}^n$ is a compact set. Then $f(K)$ is bounded and there exists $x, y \in K$ with*

$$f(x) = \inf\{f(z) \,|\, z \in K\}$$
$$f(y) = \sup\{f(z) \,|\, z \in K\}.$$

In more boiled down terms, this corollary says that a real continuous function on a compact set assumes its minimum and its maximum. This result is extremely important for the existence of optimal solutions in optimization problems.

A.8   Exercises

Exercise A.1. Use induction to prove the formula in (A.2).

Exercise A.2.

(i) Show that
$$2ab \leq a^2 + b^2$$

for $a, b \in \mathbb{R}$.

(ii) Let $x, y \in \mathbb{R}^n \setminus \{0\}$, where $x = (x_1, \ldots, x_n)$ and $y = (y_1, \ldots, y_n)$. Prove that
$$2 \frac{x_i}{|x|} \frac{y_i}{|y|} \leq \frac{x_i^2}{|x|^2} + \frac{y_i^2}{|y|^2}$$

for $i = 1, \ldots, n$.

(iii) Deduce the Cauchy-Schwarz inequality from (ii).

Exercise A.3. Show formally that $1, 2, 3, \ldots$ does not have a convergent subsequence. Can you have a convergent subsequence of a non-convergent sequence?

Exercise A.4. Prove Proposition A.9.

Exercise A.5. Let S be a subset of the rational numbers $\mathbb{Q}$, which is bounded from above. Of course this subset always has a supremum in $\mathbb{R}$. Can you give an example of such an S, where $\sup(S) \notin \mathbb{Q}$.

Exercise A.6. Let $S = \mathbb{R} \setminus \{0, 1\}$. Prove that S is not closed. What is $\overline{S}$?

Exercise A.7. Let $S_1 = \{x \in \mathbb{R} \mid 0 \leq x \leq 1\}$. What is $\text{int}(S_1)$ and ∂S_1?
Let $S_2 = \{(x, y) \in \mathbb{R}^2 \mid 0 \leq x \leq 1, \ y = 0\}$. What is $\text{int}(S_2)$ and ∂S_2?

Exercise A.8. Let $S \subseteq \mathbb{R}^n$. Show that $\text{int}(S) \subseteq S$ and $S \cup \partial S = \overline{S}$. Is ∂S contained in S?
Let $U = \mathbb{R}^n \setminus F$, where $F \subseteq \mathbb{R}^n$ is a closed set. Show that $\text{int}(U) = U$ and $\partial U \cap U = \emptyset$.

Exercise A.9. Show that
$$\big||x| - |y|\big| \leq |x - y|$$

for every $x, y \in \mathbb{R}^n$.

Exercise A.10. Give an example of a subset $S \subseteq \mathbb{R}$ and a continuous function $f : S \to \mathbb{R}$, such that $f(S)$ is not bounded.

Appendix B

Linear (in)dependence and the rank of a matrix

The concept of linear (in)dependence is often a stumbling block in introductory courses on linear algebra. When presented as a sterile definition in an abstract vector space it can be hard to grasp. I hope to show here that it is simply a fancy way of restating a quite obvious fact about solving linear equations.

B.1 Linear dependence and linear equations

You can view the equation

$$3x + 5y = 0$$

as one linear equation with two unknowns. Clearly $x = y = 0$ is a solution. But there is also a non-zero solution with $x = -5$ and $y = 3$. As one further example consider

$$2x + y - z = 0$$
$$x + y + z = 0. \tag{B.1}$$

Here we have 3 unknowns and only 2 equations and $x = 2$, $y = -3$ and $z = 1$ is a non-zero solution.

These examples display a fundamental fact about linear equations. A system

$$a_{11} x_1 + \cdots + a_{1n} x_n = 0$$
$$a_{21} x_1 + \cdots + a_{2n} x_n = 0$$
$$\vdots$$
$$a_{m1} x_1 + \cdots + a_{mn} x_n = 0$$

of linear equations always has a non-zero solution if the number of unknowns n is greater than the number n of equations i.e., $n > m$. In modern linear algebra this fact about linear equations is phrased using the concept linear dependence.

Definition B.1. A set of vectors $V = \{v_1, \ldots, v_r\} \subseteq \mathbb{R}^n$ is called linearly dependent if there exists $\lambda_1, \ldots, \lambda_r \in \mathbb{R}$ not all zero, such that

$$\lambda_1 v_1 + \cdots + \lambda_r v_r = 0.$$

Similarly V is (or $v_1, \ldots, v_r$ are) called linearly independent if V is not linearly dependent.

With this language you can restate the non-zero solution $x = 2$, $y = -3$ and $z = 1$ of (B.1) as the linear dependence

$$2 \cdot \begin{pmatrix} 2 \\ 1 \end{pmatrix} + (-3) \cdot \begin{pmatrix} 1 \\ 1 \end{pmatrix} + 1 \cdot \begin{pmatrix} -1 \\ 1 \end{pmatrix}$$

$$= \begin{pmatrix} 4 \\ 2 \end{pmatrix} + \begin{pmatrix} -3 \\ -3 \end{pmatrix} + \begin{pmatrix} -1 \\ 1 \end{pmatrix} = \begin{pmatrix} 0 \\ 0 \end{pmatrix}.$$

Let us give a simple induction proof of the fundamental fact on (homogeneous) systems of linear equations.

Theorem B.2. *The system*

$$a_{11} x_1 + \cdots + a_{1n} x_n = 0$$
$$a_{21} x_1 + \cdots + a_{2n} x_n = 0 \tag{B.2}$$
$$\vdots$$
$$a_{m1} x_1 + \cdots + a_{mn} x_n = 0$$

of linear equations always has a non-zero solution if $m < n$.

Proof.   The induction is on m – the number of equations. For $m = 1$ we have 1 linear equation

$$a_1 x_1 + \cdots + a_n x_n = 0$$

with n variables where $n > m = 1$. If $a_i = 0$ for some $i = 1, \ldots, n$ then clearly $x_i = 1$ and $x_j = 0$ for $j \neq i$ is a non-zero solution. Assume otherwise that $a_i \neq 0$ for every $i = 1, \ldots, m$. In this case $x_1 = 1, x_2 = -a_1/a_2, x_3 = \cdots = x_n = 0$ is a non-zero solution.

If every $a_{i1} = 0$ for $i = 1, \ldots, m$, then $x_1 = 1, x_2 = \cdots = x_n = 0$ is a non-zero solution in (B.2). Assume therefore that $a_{11} \neq 0$ and substitute

$$x_1 = \frac{1}{a_{11}}(-a_{12} x_2 - \cdots - a_{1n} x_n)$$

x_1 into the remaining $m - 1$ equations. This gives the following system of $m - 1$ equations in the $n - 1$ variables $x_2, \ldots, x_n$

$$\left(a_{22} - \frac{a_{21}}{a_{11}} a_{12} \right) x_2 + \cdots + \left(a_{2n} - \frac{a_{21}}{a_{11}} a_{1n} \right) x_n = 0$$
$$\vdots \qquad (\text{B.3})$$
$$\left(a_{m2} - \frac{a_{m1}}{a_{11}} a_{12} \right) x_2 + \cdots + \left(a_{mn} - \frac{a_{m1}}{a_{11}} a_{1n} \right) x_n = 0 .$$

Since $n - 1 > m - 1$, the induction assumption on m gives the existence of a non-zero solution $(a_2, \ldots, a_n)$ to (B.3). Now

$$\left(\tfrac{1}{a_{11}} (-a_{12}a_2 - \cdots - a_{1n}a_n), a_2, \ldots, a_n \right)$$

is a non-zero solution to our original system of equations. This can be checked quite explicitly (Exercise B.4). □

Theorem B.2 has the following reformulation using linear dependence.

Corollary B.3. *A set of vectors* $\{v_1, \ldots, v_n\} \subset \mathbb{R}^m$ *is linearly dependent if* $m < n$.

B.2 The rank of a matrix

Consider an $m \times n$ matrix

$$A = \begin{pmatrix} a_{11} & \cdots & a_{1n} \\ \vdots & \ddots & \vdots \\ a_{m1} & \cdots & a_{mn} \end{pmatrix}$$

with row vectors

$$A_1 = (a_{11}, \ldots, a_{1n}),$$
$$\vdots$$
$$A_m = (a_{m1}, \ldots, a_{mn}),$$

in $\mathbb{R}^n$. The *row rank*, rrk(A) of A is the maximal number of linearly independent vectors in the set of row vectors. Similarly we define the *column rank*, crk(A), to be the maximal number of linearly independent column vectors or crk$(A) =$ rrk(A^t), where A^t is the transposed matrix of A.

Example B.4. The 3×3 matrix

$$A = \begin{pmatrix} 1 & 2 & 3 \\ 4 & 5 & 6 \\ 5 & 7 & 9 \end{pmatrix}$$

has row rank < 3 as

$$\begin{pmatrix} 1 \\ 2 \\ 3 \end{pmatrix} + \begin{pmatrix} 4 \\ 5 \\ 6 \end{pmatrix} - \begin{pmatrix} 5 \\ 7 \\ 9 \end{pmatrix} = \begin{pmatrix} 0 \\ 0 \\ 0 \end{pmatrix}$$

is a linear dependence among the row vectors. In proving for example that the first two row vectors are linearly independent one looks for $\lambda, \mu \in \mathbb{R}$ with

$$\lambda \begin{pmatrix} 1 \\ 2 \\ 3 \end{pmatrix} + \mu \begin{pmatrix} 4 \\ 5 \\ 6 \end{pmatrix} = \begin{pmatrix} 0 \\ 0 \\ 0 \end{pmatrix}.$$

Therefore λ and μ must satisfy

$$\lambda + 4\mu = 0$$
$$2\lambda + 5\mu = 0,$$

which implies that $\lambda = \mu = 0$. Therefore $\mathrm{rrk}(A) = 2$. Similarly one can show that $\mathrm{crk}(A) = 2$. Surprisingly the row and column ranks always agree for a matrix.

With the help of Corollary B.3 we get

Corollary B.5. *For an $m \times n$ matrix A, we have*

$$\mathrm{rrk}(A) \leq n.$$

Lemma B.6. *Consider an $m \times n$ matrix A with rows $A_1, \ldots, A_m \in \mathbb{R}^n$. If the first s rows $A_1, \ldots, A_s$ are linearly independent and $A_1, \ldots, A_s, A_j$ are linearly dependent for every $j > s$, then for every row A_i there exists $\lambda_1, \ldots, \lambda_s \in \mathbb{R}$ such that*

$$A_i = \lambda_1 A_1 + \cdots + \lambda_s A_s \tag{B.4}$$

and $\mathrm{rrk}(A) = s$.

Proof.   Clearly (B.4) holds for $1 \leq i \leq s$. For $i > s$ we have by assumption a linear dependence

$$\mu A_i + \mu_1 A_1 + \cdots + \mu_s A_s = 0 \tag{B.5}$$

with $\mu, \mu_1, \ldots, \mu_s$ not all zero. Here we must have $\mu \neq 0$, otherwise (B.5) would be a linear dependence among $A_1, \ldots, A_s$ contradicting our assumption. Therefore

$$A_i = -(\mu_1/\mu)A_1 - \cdots - (\mu_s/\mu)A_s$$

proving (B.4). Now suppose that $A_{i_1}, \ldots, A_{i_r}$ is a linearly independent subset of the rows with $r = \mathrm{rrk}(A)$. Then there exists $\lambda_{11}, \ldots, \lambda_{rs} \in \mathbb{R}$, such that

$$A_{i_1} = \lambda_{11}A_1 + \cdots + \lambda_{1s}A_s$$
$$\vdots$$
$$A_{i_r} = \lambda_{r1}A_1 + \cdots + \lambda_{rs}A_s$$

by (B.4). A linear dependence $x_1 A_{i_1} + \cdots + x_r A_{i_r} = 0$ can then be written as

$$(\lambda_{11}x_1 + \cdots + \lambda_{r1}x_r)A_1 + \cdots + (\lambda_{1s}x_1 + \cdots + \lambda_{rs}x_r)A_s = 0,$$

which implies that

$$\lambda_{11}x_1 + \cdots + \lambda_{r1}x_r = 0$$
$$\vdots \qquad\qquad\qquad (B.6)$$
$$\lambda_{1s}x_1 + \cdots + \lambda_{rs}x_r = 0$$

by the linear independence of $A_1, \ldots, A_s$. If $s < r$, then (B.6) has a nonzero solution $x_1, \ldots, x_r$ by Theorem B.2. This would contradict the linear independence of $A_{i_1}, \ldots, A_{i_r}$. Therefore $r \leq s$. Since $r \geq s$ by definition, it follows that $r = s$. $\qquad\square$

Corollary B.7. *For any $m \times n$ matrix A, we have*

$$\mathrm{rrk}(A) = \mathrm{crk}(A).$$

Proof. Suppose that $\mathrm{rrk}(A) = r$. Then we may assume that the first r rows $A_1, \ldots, A_r$ are linearly independent and that $A_1, \ldots, A_r, A_j$ are linearly dependent for every $j > r$. Now (B.4) implies that

$$Ax = (A_1 x, \ldots, A_r x, \ldots, A_m x) = 0$$

if and only if $(A_1 x, \ldots, A_r x) = 0$ for $x \in \mathbb{R}^n$. This shows that the column rank of A agrees with the column rank of the $r \times n$ matrix with rows $A_1, \ldots, A_r$. However the column rank of this matrix is $\leq r$ by Corollary B.5. Therefore $\mathrm{rrk}(A) \geq \mathrm{crk}(A)$. But $\mathrm{crk}(A) = \mathrm{rrk}(A^t) \geq \mathrm{crk}(A^t) = \mathrm{rrk}(A)$. Therefore $\mathrm{rrk}(A) = \mathrm{crk}(A)$. $\qquad\square$

Definition B.8. The rank of a matrix A is defined as

$$\mathrm{rk}(A) := \mathrm{rrk}(A) = \mathrm{crk}(A).$$

Now we are ready for our main result tailored to the application of linear algebra to convex cones and polyhedra.

Theorem B.9. *Let A be an $m \times n$ matrix of rank n. Suppose that $r < n$ and that the first r rows $A_1, \ldots, A_r \in \mathbb{R}^n$ of A are linearly independent.*

(1) *There exists $r < j \le n$, such that $A_1, \ldots, A_r, A_j$ are linearly independent.*

(2) *With the notation from (1) there exists $x \in \mathbb{R}^n$ with $A_1 x = \cdots = A_r x = 0$ and $A_j x \ne 0$.*

(3) *If $r = n - 1$, then any $x \in \mathbb{R}^n$ with $A_1 x = \cdots = A_r x = 0$ is uniquely determined up to multiplication by a constant.*

Proof.   The proof of (1) follows from Lemma B.6. Consider the matrix A' with rows $A_1, \ldots, A_r$ and the matrix A'' which is A' extended by the row A_j. Then $\mathrm{crk}(A'') = \mathrm{rrk}(A'') = \mathrm{rrk}(A')+1 = \mathrm{crk}(A')+1 =: s$. Therefore A'' has s linearly independent columns. The corresponding columns in A' have to be linearly dependent i.e., there exists $x \in \mathbb{R}^n$ with $A'x = 0$ but $A''x \ne 0$. This proves (2). By (2) there exists $x \in \mathbb{R}^n \setminus \{0\}$ such that $A_1 x = \cdots = A_{n-1} x = 0$. The vectors $A_1, \ldots, A_{n-1}$ and x are linearly independent:

$$z = \lambda_1 A_1 + \cdots + \lambda_{n-1} A_{n-1} + \lambda x^t = 0 \qquad (B.7)$$

implies $zx = \lambda |x|^2 = 0$. Therefore $\lambda = 0$ and $\lambda_1 = \cdots = \lambda_{n-1} = 0$, since $A_1, \ldots, A_{n-1}$ are linearly independent. By Lemma B.6 and Corollary B.5 any vector in $\mathbb{R}^n$ is a linear combination of $A_1, \ldots, A_{n-1}$ and x. Thus if x_1 is a vector with $A_1 x_1 = \cdots = A_{n-1} x_1 = 0$, then

$$x_1^t = y + \mu x^t \qquad (B.8)$$

for $y = \mu_1 A_1 + \cdots + \mu_{n-1} A_{n-1}$ and $\mu_1, \ldots, \mu_{n-1}, \mu \in \mathbb{R}$. Therefore $0 = y^t x_1^t = |y|^2 + \mu y^t x^t = |y|^2$ and it follows that $x_1 = \mu x$. $\square$

B.3   Exercises

Exercise B.1. Find $\lambda_1, \lambda_2, \lambda_3 \in \mathbb{R}$ not all 0 with

$$\lambda_1 \begin{pmatrix} 1 \\ 2 \end{pmatrix} + \lambda_2 \begin{pmatrix} 3 \\ 4 \end{pmatrix} + \lambda_3 \begin{pmatrix} 5 \\ 6 \end{pmatrix} = \begin{pmatrix} 0 \\ 0 \end{pmatrix}.$$

Exercise B.2. Show that a non-zero solution (x, y, z) to (B.1) must have $x \neq 0, y \neq 0$ and $z \neq 0$. Is it possible to find $\lambda_1, \lambda_2, \lambda_3$ in Exercise B.1, where one of λ_1, λ_2 or λ_3 is 0?

Exercise B.3. Can you find a non-zero solution to

$$x + y + z = 0$$
$$x - y + z = 0,$$

where

(i) $x = 0$?
(ii) $y = 0$?
(iii) $z = 0$?
(iv) What can you say in general about a system

$$ax + by + cz = 0$$
$$a'x + b'y + c'z = 0$$

of linear equations in x, y and z, where a non-zero solution always has $x \neq 0, y \neq 0$ and $z \neq 0$?

Exercise B.4. Check carefully that

$$\left(\tfrac{1}{a_{11}}(-a_{12}a_2 - \cdots - a_{1n}a_n), a_2, \ldots, a_n \right)$$

really is a non-zero solution to (B.2) in the proof of Theorem B.2.

Exercise B.5. Compute the rank of the matrix

$$\begin{pmatrix} 1 & 2 & 3 \\ 4 & 5 & 6 \\ 14 & 19 & 24 \\ 6 & 9 & 12 \end{pmatrix}.$$

Bibliography

Almkvist, G. and Berndt, B. (1988). Gauss, Landen, Ramanujan, the Arithmetic-Geometric Mean, Ellipses, π, and the Ladies Diary, *The American Mathematical Monthly* **95**, pp. 585–608.

Barvinok, A. (2002). *A Course in Convexity, Graduate Studies in Mathematics*, Vol. 54 (American Mathematical Society, Providence, RI).

Bazaraa, M., Sherali, H. and Shetty, C. (2006). *Nonlinear Programming: Theory and Algorithms* (Wiley-Interscience).

Boyd, S. and Vandenberghe, L. (2004). *Convex optimization* (Cambridge University Press, Cambridge).

Brøndsted, A. (1983). *An introduction to convex polytopes, Graduate Texts in Mathematics*, Vol. 90 (Springer-Verlag, New York).

Courant, R. and Robbins, H. (1941). *What Is Mathematics?* (Oxford University Press, New York).

Dantzig, G. B. (1963). *Linear programming and extensions* (Princeton University Press, Princeton, N.J.).

de Loera, J. (2005). The many aspects of counting lattice points in polytopes, *Mathematische Semesterberichte* **52**, 2, pp. 175–195.

Feller, W. (1957). *An introduction to probability theory and its applications. Vol. I* (John Wiley and Sons, Inc., New York), 2nd ed.

Feynman, R., Leighton, R. and Sands, M. (2006). *The Feynman Lectures on Physics* (Addison-Wesley Publishing Company).

Fourier, J. (1826). Solution d'une question perticulière du calcul des inégalités, *Nouveau Bulletin des Sciences par la Société philomatique de Paris* , pp. 317–319.

Fukuda, K. and Prodon, A. (1996). Double description method revisited, in *Combinatorics and computer science (Brest, 1995), Lecture Notes in Comput. Sci.*, Vol. 1120 (Springer, Berlin), pp. 91–111.

Gordan, P. (1873). Ueber die Auflösung linearer Gleichungen mit reellen Coefficienten, *Math. Ann.* **6**, 1, pp. 23–28.

Graham, R. (1972). An efficient algorithm for determining the convex hull of a finite planar set, *Information Processing Letters* **1**, pp. 132–133.

Grcar, J. F. (2011). Mathematicians of Gaussian elimination, *Notices Amer.*

Math. Soc. **58**, 6, pp. 782–792.

Iversen, B. (1988). *Reelle funktioner af flere variable* (Aarhus Universitetsforlag).

Jacobi, C. G. (1829). *Fundamenta nova theoriae functionum ellipticarum* (Königsberg).

Jensen, J. L. W. V. (1905). Om konvekse Funktioner og Uligheder imellem Middelværdier, *Nyt Tidskrift for Mathematik* **B. 16**, pp. 49–68.

Jensen, J. L. W. V. (1906). Sur les fonctions convexes et les inégalités entre les valeurs moyennes, *Acta Math.* **30**, pp. 175–193.

Karush, W. (1939). *Minima of Functions of Several Variables with Inequalities as Side Constraints*, Master's thesis, University of Chicago.

Kjeldsen, T. H. (2008). From measuring tool to geometrical object: Minkowski's development of the concept of convex bodies, *Arch. Hist. Exact Sci.* **62**, 1, pp. 59–89.

Knuth, D. E. (1981). A permanent inequality, *The American Mathematical Monthly* **88**, 10, pp. 731–740.

Kuhn, H. W. and Tucker, A. W. (1951). Nonlinear programming, in *Proceedings of the Second Berkeley Symposium on Mathematical Statistics and Probability, 1950* (University of California Press, Berkeley and Los Angeles), pp. 481–492.

Motzkin, T. S., Raiffa, H., Thompson, G. L. and Thrall, R. M. (1953). The double description method, in *Contributions to the theory of games, vol. 2*, Annals of Mathematics Studies, no. 28 (Princeton University Press, Princeton, N. J.), pp. 51–73.

Rockafellar (1970). *Convex Analysis* (Princeton University Press).

Schrijver, A. (1986). *Theory of linear and integer programming*, Wiley-Interscience Series in Discrete Mathematics (John Wiley & Sons Ltd., Chichester), a Wiley-Interscience Publication.

Slater, M. (1950). Lagrange multipliers revisited: a contribution to non-linear programming, Discussion Paper Math. 403, Cowles Commission.

Spielman, D. and Teng, S.-H. (2001). Smoothed analysis of algorithms: why the simplex algorithm usually takes polynomial time, in *Proceedings of the Thirty-Third Annual ACM Symposium on Theory of Computing* (ACM, New York), pp. 296–305 (electronic).

Stoer, J. and Bulirsch, R. (1980). *Introduction to numerical analysis* (Springer-Verlag, New York), translated from the German by R. Bartels, W. Gautschi and C. Witzgall.

Sylvester, J. J. (1857). A question in the geometry of situation, *Quarterly Journal of Pure and Applied Mathematics* **1**, p. 79.

Toft, B. (1988). *Algoritmisk kombinatorik: et datalogisk orienteret emne i matematik* (Matematiklærerforeningen and Odense Universitet. Matematisk Institut).

von Neumann, J. and Morgenstern, O. (1944). *Theory of Games and Economic Behavior* (Princeton University Press, Princeton, New Jersey).

Weyl, H. (1935). Elementare Theorie der konvexen Polyeder, *Commentarii Mathematici Helvetici* **7**, pp. 290–306.

Wright, M. (2005). The interior-point revolution in optimization: history, recent developments and lasting consequences. *Bull. Amer. Math. Soc.* **42**, pp. 39–56.

Ziegler, G. M. (1995). *Lectures on polytopes, Graduate Texts in Mathematics*, Vol. 152 (Springer-Verlag, New York).

Index